T0209498

essentials

Springer essentials

Springer essentials provide up-to-date knowledge in a concentrated form. They aim to deliver the essence of what counts as "state-of-the-art" in the current academic discussion or in practice. With their quick, uncomplicated and comprehensible information, *essentials* provide:

- an introduction to a current issue within your field of expertise
- an introduction to a new topic of interest
- an insight, in order to be able to join in the discussion on a particular topic

Available in electronic and printed format, the books present expert knowledge from Springer specialist authors in a compact form. They are particularly suitable for use as eBooks on tablet PCs, eBook readers and smartphones. *Springer essentials* form modules of knowledge from the areas economics, social sciences and humanities, technology and natural sciences, as well as from medicine, psychology and health professions, written by renowned Springer-authors across many disciplines.

Jonas Michael Wilhelm Westphal

The Sustainability of Hemp

An Overview of Product and Use in Business

Jonas Michael Wilhelm Westphal
Grönwohld, Germany

ISSN 2197-6708   ISSN 2197-6716   (electronic)
essentials
ISSN 2731-3107   ISSN 2731-3115   (electronic)
Springer essentials
ISBN 978-3-658-41818-2   ISBN 978-3-658-41819-9   (eBook)
https://doi.org/10.1007/978-3-658-41819-9

This Springer imprint is published by the registered company Springer Fachmedien Wiesbaden
GmbH, part of Springer Nature.
The registered company address is: Abraham-Lincoln-Str. 46, 65189 Wiesbaden, Germany

What You Can Find in This *essential*

- An overview of the UN's sustainability policy (Agenda 2030) and the Federal Republic
- An insight into the botany, economics and legal situation of hemp
- How hemp could serve environmental and climate protection within the 17 sustainability goals (SDG)
- How hemp (THC) is used as medicine
- How the benefits of industrial hemp could be politically implemented economically and ecologically through legalization, education and research

Foreword

For me, who has been since 40 years fighting for the legalization of *cannabis*,[1] it was a real "aha" moment to read how helpful it would be if we were to use the "positive" sides of the same plant, namely hemp, or *Cannabis sativa L*, in one of the most important tasks of our time, namely to stop the destruction of our world by living *sustainably*.

And that was a double "aha" experience of a very fundamental kind. First, to find out how much our drug-fogged view always only had the negative sides of this plant in mind; whether this, as is still customary, should be the youth and brain-corrupting *THC* drug, or, as we "anti-prohibitionists" meant, the risks of the repression used against it. With which these two "drug-political" opponents barricaded themselves so mutually in a thought-prison that they could only with difficulty and very limitedly recognize this drug, the natural *THC*, as a medicine under the supervision of the Federal Opium Agency. And that one—unlike, for example, the "poisonous" potato plant[2]—even banished the 'industrial hemp' in these very prison gardens: first, to completely ban it for a short time, from 1982 to 1996, and then to anchor it in the *Narcotics Act (BtMG)* under even more conditions: if it contains 'less than 0.2% *THC*' or is to be used as a 'barrier strip in beet breeding', in order then to be grown preferably only by agricultural companies, which are to serve the 'old-age security of farmers'. An 'aha' experience that once again made me aware of how one-sidedly we perceive our reality with a good conscience, how often we also—quite without any bad intention—simply

---

[1] As author of the book *Drogenelend* (Campus-Verlag 1982) as well as co-founder of the anti-prohibition association *Schildower Kreis* (https://schildower-kreis.de).

[2] https://gizbonn.de/giftzentrale-bonn/pflanzen/kartoffel

ignore everything or at least cut it to size if it does not fit into this framework, into this drug-*frame*.

If you read here in this publication about the exemplary possibilities of using this industrial hemp 'sustainably'—in agriculture for soil improvement and biodiversity, as a durable fibre instead of cotton or as 'biological plastic', or even in the construction industry as insulation material and hemp lime, which can be recycled as part of *urban mining*—then you are always amazed how much we instead destroy our environment with monocultures of maize, pesticides, cement and plastic, although we were already warned about the finiteness of our resources half a century ago, that is, two generations ago, by the *Club of Rome*[3]. A 'after us the deluge' attitude that we actually only had to be made aware of by the *'Friday for Future'* youth.

Two very strange blockades—in thinking and in action—which did not fall from the sky or which arose 'naturally' from our genes, as convenient as this may be for us as an excuse not to do anything about it anyway. No, two blockades that have been produced, propagated and maintained 'interest-politically' by people since the middle of the last century, in the wake of the experiences of the Second World War: the US-American *'War on Drugs'*[4] on the one hand and the mass production of plastic on the other.[5] Two powerful actor groups: The state, that is, the ministries, parties, police, judges, along with the therapists and warning psychiatrists interested in "drugs" on the one side. And the economy, the plastic industry, the packaging company and the industrial agriculture oriented towards factory farming and large farms on the other side. Powerful stakeholders who in turn secure themselves internationally: At the European level, for example, in the "Varieties Catalogue of Certified Seed" jointly adopted by the European Parliament and Council, which then landed in the German "BtMG"; as well as in the international conventions, such as the "Single Convention on Narcotic" *Drugs* of 1961,[6] which requires mandatory the criminalization of cannabis; and of course

---

[3] Cf. Jared Diamond (2006): Collapse. Why societies survive or perish. Fischer Taschenbuch Verlag.

[4] https://de.wikipedia.org/wiki/War_on_Drugs

[5] https://www.bund.net/fileadmin/user_upload_bund/publikationen/chemie/chemie_plastikatlas_2019.pdf

[6] https://de.wikipedia.org/wiki/Einheitsabkommen_über_die_Betäubungsmittel

the large international economic complexes, which, à la *McDonald*, recently took legal action against the city of Tübingen's attempt to introduce a plastic tax.[7]

But it seems that in recent years the tide is also turning here. On the one hand, an internationally highly staffed *World Commission on Drug Policy* has been fighting for the decriminalization of drug policy since 2011, while on the other hand, the *UN Sustainability Reports* underlying this work open up a new perspective. On a national level, the decriminalization of *cannabis* is gaining momentum again, driven by the USA, also as a project othe German government. *THC* as a medicine and *CBD* as a food and cosmetic agent are breaking new ground in the still dominant drug-minded prison. And the first experiments and examples described in this book show the possibility of using hemp itself in the fight against CO_2 reduction. Possibilities that are urgently dependent on research that is still almost completely lacking, but promising. That is why the "state", in particular the relevant federal and state ministries, are called upon to support these opportunities financially, both in research and in the area of subsidies. For which they could and should also rely on this writing with its rich source references.

Mai 2022   Prof. Stephan Quensel

---

[7] https://www.change.org/p/mario-federico-vorstandsvorsitzender-von-mcdonald-s-deutschland-umdenkenmcd-mcdonald-s-muss-klage-gegen-verpackungssteuer-in-tübingen-zurückziehen

Contents

List of Figures

Introduction—What It's About

The general view of Cannabis Sativa is becoming increasingly liberal; the voices of legalization are becoming more political and expressive; the legal cannabis economy is rapidly expanding nationally and internationally. And the demand for sustainability is coming up internationally, as in Germany, increasingly in the discussion, especially in terms of climate and ecology. This work examines how the use of the plant Cannabis Sativa can contribute to achieving the goals of the UN Sustainability Report, the so-called Agenda 2030 and its 17 SDGs (Sustainable Development Goals Sustainable Development Goals) and how this is being or could be realized in the Federal Republic of Germany.

In contrast to the clearly ideologically colored legalization debate[1], an independent cannabis economy has developed since the 1980s, which initially drove the consumption-oriented trade and use as a coffee shop, club and self-cultivation idea in order to then to emphasize and practically implement the traditionally flourishing use and technical use of hemp. It relies on the plant properties of the hemp plant: fibers, seeds, as well as its specific (drug) pharmacology: medicines, cosmetics. Two 'positive', economic-ecological aspects for a comprehensive cannabis discussion, which I will limit myself to in this work, without going into the still weighty traditional legalization arguments or the recently beginning CBD discussion, both of which are briefly mentioned in the following section.

So in the current cannabis debate, the traditional legalization-illegalization perspective still dominates, in which one side[2]—ethnologically-culturally and

---

[1] See, for example: Bureg/BÜNDNIS 90/DIE GRÜNEN (2020).

[2] Global Commission on Drug Policy (2021); Klein and Stothard (2018).

J. M. W. Westphal, *The Sustainability of Hemp*, Springer essentials, https://doi.org/10.1007/978-3-658-41819-9_1

historically-sociologically—emphasizes the mostly subcultural recreational consumption as 'positive' aspects of cannabis, referring to the basic rights of the Basic Law ('Grundgesetz') and the UN Human Rights Convention and emphasizing the negative aspects of repression.[3] While the other, psychiatrically-criminal law-dominated side emphasizes only its harmful effects on individual and social health in order to restrict or even eradicate the use of cannabis as an 'illegalized' drug threatened with penalties.[4]

The discussion about the use of pure CBD (Cannabidiol),[5] which, unlike THC, has no psychoactive effect, is not discussed in more detail. Of the three possible uses—as food, medicine and cosmetics—the food sector is currently controversial,[6] the medical use is only approved for two agents, while CBD is freely accessible as a cosmetic, especially on the Internet,[7] where hemp oils and capsules[8] are also advertised as remedies in practice.

---

[3] Böllinger (2016).

[4] Jähnert (2021); Drug Commissioner (2021).

[5] For discussion: Knodt (2021); Wurth (2020).

[6] Administrative Court Berlin (2021).

[7] Steinort (2021).

[8] Stiftung Warentest (2021); Short version: Stiftung Warentest (2021a).

Sustainability

2

The idea of having to operate sustainably on a global scale only gained importance at the beginning of this century. Building on the long-ignored report of the Club of Rome *'The Limits of Growth'* (1972), which pointed to the finite limits of world resources, the World Commission on Environment and Development published its report in 1987 under the title *"Our Common Future. The Brundtland Report of the World Commission on Environment and Development"*, which for the first time provided a concrete definition of the concept of sustainable development: *"Sustainable development is development that satisfies the needs of the present without risking that future generations cannot satisfy their needs."*[1] After several follow-up conferences of the UN: 1992 in Rio de Janeiro, 2000 in New York (Millennium Summit), 2002 in Johannesburg and 2012 again in Rio de Janeiro, the UN adopted the resolution *'Agenda 2030'* for sustainable development with 17 goals and a total of 169 sub-goals (Sustainable goals: SDGs), which Ban Ki-moon, the then Secretary-General of the United Nations, summarized as follows: *"The agenda not only aims to eradicate extreme poverty, but also to integrate the three dimensions of sustainable development—the economic, social and ecological dimensions—into a comprehensive global vision."*[2]

In the report of the follow-up conference of 2020, the current Secretary-General of the UN, António Guterres, emphasizes these goals in his foreword[3] with particular emphasis on the now increasingly coming into focus climate

---

[1] Brundtland Report (1987).

[2] United Nations (2016).

[3] United Nations (2020).

J. M. W. Westphal, *The Sustainability of Hemp*, Springer essentials, https://doi.org/10.1007/978-3-658-41819-9_2

crisis:[4] *"The Agenda 2030 for Sustainable Development was launched in 2015 to end poverty and put the world on a path of peace, prosperity and opportunity for all on a healthy planet. To do this, we must win the race against climate change, take decisive action against poverty and inequality, empower all women and girls to true self-determination and create inclusive and equitable societies everywhere."* But he also warns: *"The present report, based on the latest data, shows that progress was uneven and inadequate to achieve the goals by 2030, even before the COVID-19 pandemic. [...]* *"Now, the COVID-19 pandemic is threatening lives and livelihoods with an unprecedented health, economic and social crisis, making it even more difficult to achieve the goals."* A warning that is shockingly underlined by the recent Sixth Assessment Report *Climate Change 2021: The Physical Science Basis, the Working Group I* (06.08. 2021)[5].

In Germany, the institutionalization of this demand for sustainability began,[6] the *"hardly known to the general public [...]"*,[7] 2001 with the establishment of a Sustainability Council convened by the Federal Government *Council for Sustainable Development* (RNE) and since 2004 by a *Parliamentary Advisory Council for Sustainable Development* (PBnE), which *"accompanies the national and European sustainability strategy. It also checks the sustainability impact assessment of laws."*[8] Coordinated by a State Secretaries Committee led by the Federal Chancellery *State Secretaries Committee* (St-Ausschuss), *"in which all departments are represented at the level of the technically responsible sub-department managers"*[9], advised by a *German Solution Network for Sustainable Development* (SDSN Germany), in which *"leading German knowledge organizations and partners from business and society have been working together since 2014"*,[10] and on the basis of a complex 'societal dialogue 2019/2020',[11] the Federal Government revised its own sustainability goals in line with the latest UN resolution of 2020, which are also increasingly driven by the climate protection idea:[12]

---

[4] Paris Climate Agreement (2015).

[5] IPCC (2021); IPCC (2022); Federal Environment Agency (2022).

[6] DNS Development (2021) and there the Fig. Institutions.

[7] United Nations (2016, p. 232).

[8] DNS (2017, p. 14).

[9] DNS (2017, p. 249).

[10] DNS Draft (2021, p. 37).

[11] DNS Draft (2021, p. 8).

[12] DNS Development (2021).

Judgment of the Federal Constitutional Court of 29.04.2021[13] as well as new version of the Climate Protection Act.[14]

---

[13] Traufetter (2021).
[14] Klimapakt Deutschland (2021); Climate Protection Act (2021).

Cultural Plant Hemp

3

The ancient crop hemp,[1] which was also prohibited as industrial hemp in Germany from 1982 to 1996,[2] can be used from an ecological point of view as a 'natural product', i.e. without recourse to further raw materials and without residues that can be recycled without waste, which is why the Federal Ministry of Food and Agriculture can write:[3] *"Hemp—scientifically correctly called Cannabis sativa—was used in China more than 10,000 years ago. Different hemp varieties can be grown in almost all climatic zones of the world—in subtropical South America as well as in the harsh climate of Siberia. The one-year-old plants grow up to four meters high in six months and are extremely versatile. The entire plant can be used from the stem to the flower to the seeds."*

Hemp can also be grown under Central European climatic conditions, both on bog soils and on fallow land, and even on so-called 'inferior' and water-poor soils, as the roots reach up to two meters into the ground in search of groundwater.[4] It grows as fiber hemp optimally up to 2 to 2.5 m[5] and as seed hemp, similar to maize, within 100 days up to a height of 4 to 5 m. It requires hardly any plant protection products (PSM: insecticides, pesticides) and moderate nitrogen fertilizer. This is what the Federal Government's answer of 4.7.2019 to the

---

[1] Herer and Bröckers (1994, pp. 115–198).
[2] Bòcsa and Karus (1997, p. 21).
[3] Klöckner (2018, p. 21).
[4] Bòcsa and Karus (1997, pp. 26 ff.)
[5] Herer and Bröckers (1994, pp. 299–372).

Left's request says:[6]*"Expenditures for plant protection products when growing industrial hemp (both herbicides and fungicides or insecticides) are currently very low or not necessary. Hemp could therefore be a very interesting crop rotation element for agricultural practice in the future."* It improves soil quality: *"many advantages, such as the good pre-crop effect, the positive effects on root penetration of the soils and the soil structure, the relatively low intensity level in fertilization and plant protection, the high soil/erosion protection and the low risk of nutrient leaching into groundwater and surface water." "The cultivation of summer crops, especially leafy crops, has a positive effect on the loosening of cereal-intensive crop rotations."*[7] But winter crops also recommend themselves as a good catch crop, as Susanne Richter can prove in her agricultural-empirical dissertation (2018): *"The cultivation of fiber hemp (Cannabis sativa L.) as a winter catch crop offers farmers an additional value creation and increases the production of the natural fiber raw material, which is increasingly used in innovative materials."*[8]

Hemp is bred in three clearly distinct subtypes: *"Cannabis is divided mainly into three phenotypes: phenotype I (drug-type), with Δ9-Tetrahydrocannabinol (THC) > 0.5% and cannabidiol (CBD) < 0.5%; phenotype II (intermediate type), with CBD as the major cannabinoid but with THC also present at various concentrations; and phenotype III (fiber-type or hemp), with especially low THC content."*[9] In my work I combine type II and III: as 'utility hemp' without the pharmaceutical 'drug' component THC (Delta-9-THC), but with CBD (Cannabidiol), for use as 'industrial hemp'.

3.1 Utility hemp (Industrial hemp)

This utility hemp is used for its wood-like stalks and its fibers, whose 'primary' fiber bundles can be up to two meters long. Its admissibility is exempted from the general cannabis ban in the Narcotics Act (BtMG)§ 19, Annex 1, if the plants are listed in the *"Common Catalogue of Varieties of Agricultural Plant Species*[10] *or*

---

[6] Bureg/DIE LINKE (2019).

[7] Bòcsa and Karus (1997).

[8] Richter (2018).

[9] Suman et al. (2017, p. 81).

[10] BLE Variety List (2022).

their content of tetrahydrocannabinol does not exceed 0.2% and the traffic with them (except for cultivation) is exclusively for commercial or scientific purposes that exclude abuse for intoxication purposes."[11] A limit that is below the European level of 0.3%,[12] and which can quickly lead to the destruction of the entire field due to natural fluctuations.

According to the answer of the Federal Government of 04.07.2019 to the request of the Left Dr. Kirsten Tackmann et al.[13]on 2148 ha 3651.6 t of utility hemp were cultivated in Germany in 2018 and 6158.8 t mainly imported from Canada, Holland and Portugal; in the EU a total of 26,900 ha were cultivated in 2018, with the focus with last 16,500 ha in France.

The 'potentials' of utility hemp, which are also politically controversial, become particularly clear in the joint motion of the Die Linke and Die Grünen factions of 14.1.2021, which criticize above all its link to the BtMG regulations: *"At the same time [to the 6 guidelines of the Ackerbaustrategie 2035], domestic utility hemp cultivation can also contribute to achieving the 17 Sustainable Development Goals of the UN and thus to securing sustainable development on economic, social and ecological levels."*[14]

3.2   Medical Hemp

Hemp of phenotype I is used as 'medical hemp' to obtain THC, which was first discovered by Raphael Mechoulam in 1964 and CBD, which has only recently gained interest, while most of the other '120 phytocannabinoids', such as *"Tetrahydrocannabivarin (THCV), Cannabichromene (CBC), Cannabigerol (CBG) and Cannabinol (CBN)"*[15] have so far been hardly studied. It is therefore subject to the BtMG, which I do not go into in more detail, but its cultivation has been approved and regulated by the Cannabis Agency at the Federal Opium Agency since 19.03.2017 in § 19 BtMG, since 2019, with *"the start-up Cansativa"*[16]

---

[11] BtMG § 1, Annex I.
[12] HEMPToday (2020).
[13] Bureg/DIE LINKE (2019).
[14] Bureg/DIE LINKE/Bündnis 90/DIE GRÜNEN (2021, p. 2).
[15] Suman et al. (2017, S. 84).
[16] Endris (2020); CANSATIVA (2021).

taking over the necessary logistics as an exclusive[17] partner.[18] Since late summer 2021, Germany has also been participating in this market with 10,400 kg allowed in 4 years, that is 2600 kg annually,[19] which are now also being grown on selected fields in Germany.[20] While so far these cannabis flowers were obtained from standardized operations in Canada, the Netherlands, Portugal, and more recently from Lesotho/South Africa and Jamaica[21] (with predetermined THC and CBD content), of which *"more than nine tons were imported to Germany in 2020— 37% more than in the previous year."*[22]

3.3   Seeds[23]

The seed can be obtained either from the separately bred seed hemp or from the dual-use hemp.[24] It contains no THC and is mainly used for hemp seed flour and hemp seed oil. According to the same legal amendment, it is not subject to the BtMG, *"provided it is not intended for unauthorized cultivation."* Hemp seeds are cholesterol and gluten free, as well as free of residues, *"they contain large amounts of high-quality and easily digestible protein, up to 25 g per 100 g of hemp seeds."*[25] They consist of more than 30% oil, 25% proteins (edestin, albumin), dietary fiber, vitamins, minerals and large amounts of all *essential* amino acids, their oil contains 80% polyunsaturated fatty acids and a high dosage of two *essential* fatty acids[26].

---

[17] BfArM (2020); BfArm (2022).
[18] Telgheder (2021).
[19] BfArM (2021).
[20] Telgheder (2021).
[21] Boedefeld (2021); Cantourage (2021).
[22] Bureg/BÜNDNIS 90/DIE GRÜNEN, (2020).
[23] Unkart (2020).
[24] Bòcsa and Karus (1997 p. 130, 156).
[25] We live sustainably (o. J.)
[26] Callaway (2004 pp. 65–72).

UN Goals (SDG) of Sustainable Development

4

If one wants to assign the sustainability potential of hemp to the individual SDGs, one encounters a double difficulty: On the one hand, the 17 SDGs have so far been developed very generally according to the original UN ambitions: No poverty (SDG 1), No hunger (SDG 2), Gender Equality (SDG 5), Less Inequality (SDG 10), Peace Justice and Strong Institutions (SDG 16) and Partnerships for Achieving the Goals (SDG 17). On the other hand, certain projects fulfill several SDGs to a different extent, so that the Federal Government in its paper 'German Sustainability Strategy Development 2021' states, for example, in the explanation of SDG 6 (Groundwater): *"The achievement of these goals is also of great importance for other SDGs, in particular Health (SDG 3), Gender Equality (SDG 5), Energy (SDG 7), Economy and Industry (SDGs 8, 9), Cities (SDG 11) and Nutrition and Agriculture and Forestry (SDGs 2, 15)."*[1]

Because of these correlations with each other, the Federal Government in its most recent 'Sustainability Strategy 2021' has formed the following summarizing *"six transformation areas"* from SDG 3 to 15: 1. Human Well-being and Capabilities, Social Justice (SDG 3, 4, 5, 8, 9, 10). 2. Energy Transition and Climate Protection (SDG 7 and 13), 3. Circular Economy (SDG 8, 9, 12), 4. Sustainable Building and Transport Transition (SDG 7, 8, 9, 11, 12, 13), 5. Sustainable Agricultural and Food Systems (SDG 2, 3, 8, 12, 13) and 6. Pollutant-free Environment (SDG 6, 8, 9, 14, 15).[2] Previously treated in detail, but not included in the transformation areas, were the goals: No poverty (SDG 1), Peace Justice and

---

[1] DNS—Dialogue Version (2021, p. 148).

[2] DNS—Dialogue Version (2021, pp. 49–60).

Strong Institutions (SDG 16) and Partnerships Achieving the Goals (SDG 17). But these goals are also included in Chapter C, in which the 17 goals are presented individually with their statistical indicators.

Of course, the special sustainability potential of hemp/cannabis does not cover all SDGs equally, nor can it be limited to certain SDGs alone. It therefore seems sensible to also limit the sustainability opportunities lying in hemp to the typical focuses for him, for which, after SDG 2 (Hunger) and SDG 3 (Health), SDG 6 and 14 (clean water, oceans), and, in reference to the 4th transformation area, the SDG 11 (Sustainable Cities and Communities) for the alternative hemp building materials and 12 (Innovation, Consumption and Production) for the other alternative hemp materials, offer themselves, which in their way open up new sources of income and employment and thus also fulfill SDG 1 (Poverty) and SDG 8 and 9 (Work and Industry) respectively. The two heavyweights SDG 13 (Climate) and SDG 15 (Agriculture) are each treated separately. On the occasion of the remaining two SDGs, the work of criminal justice is briefly mentioned under goal 16 (Peace, Justice, Strong Institutions). The need for research is mentioned under goal 4 (Education) and the role of the actors is discussed under SDG 17 (Partnerships for Achieving the Goals). No application case could be found for SDG 5 (Gender Equality).

4.1   SDG 2: No Hunger

Of course, cannabis can hardly contribute anything to solving the global hunger problem, which was generally formulated under SDG 2 in the 'German Sustainability Strategy 2021': *"Sustainable, resilient, innovative and productive agriculture is the key to global food security."*[3] Nor will one want to include the cannabis THC consumption of the *"around 11.1% of young people aged 12 to 17"* [who consumed cannabis in Germany in 2019] *"at least once in the last 12 months"*[4] under this SDG. But this flower also proves itself to be a real 'superfood' with a balanced nutrient content through its rich content of cannabinoids. It can be drunk as tea or processed into oils or butter. It can also expand the local food palette.[5] The seeds, which can be used like flax seeds,[6] are known, for example in

---

[3] DNS—Dialog version (2021, p. 138).

[4] Statista (2020).

[5] EIHA (2020); 23 Hemp Recipes: in: kochbar (2022); Gebhardt (2016).

[6] Unkart (2020).

the morning muesli, but flour or bran can also be obtained from the seeds, while hemp cake is produced as a by-product in the production of hemp seed oil, which can be used as animal feed, but is also appreciated by vegetarians and vegans as an important plant-based source of protein,[7] especially since hemp contains vitamin B 2 like in animal products.[8]

The nutty hemp oil[9] obtained from the seeds is *"one of the best cooking oils"*, which contains the essential omega 6 fatty acid (gamma linolenic acid GLA), which is currently obtained from evening primrose and borage,[10] and should not be confused with the CBD oil hype—*"An extract from THC-low/free but CBD-rich hemp flowers dissolved in a base oil (olive oil or hemp oil)"*[11]

Attempts also showed that the omega 3 and omega 6 content of eggs increased significantly when hemp was mixed into animal feed, such as in chickens[12], or that feeding hemp to cows drastically improved the production and quality of meat and milk.[13]

4.2 SDG 3: Health and Well-being

In the original UN report of 2016[14] it was said very generally about Goal 3: *"To achieve Goal 3, it is necessary to improve reproductive health and the health of mothers and children, to end HIV/AIDS, malaria, tuberculosis and neglected tropical diseases, to reduce non-communicable and environmental diseases and to ensure health care and access to safe, affordable and effective medicines and vaccines for all."* While the federal government mainly focuses on prevention and digitization in its "further developed German Sustainability Strategy 2021". It also wants, as a result of the Corona pandemic, to further strengthen its own health system and above all in cooperation with the WHO to be effective in the interest of the Third World.[15]

---

[7] We live sustainably (o. J.).

[8] USDA (2019).

[9] Herer and Bröckers (1994, pp. 338–347).

[10] Rehberg (2022); Bòcsa and Karus (1997, p. 161).

[11] Rehberg (2022).

[12] Neijat (2015).

[13] Karlsson et al. (2010).

[14] United Nations (2016).

[15] DNS Development (2021, p. 150 ff.).

But here one finds little about the need for a medical-pharmaceutical development of this SDG 3, although the current worldwide Corona epidemic, which can also deeply color the UN report 2020, shows how much the extremely rapid vaccine development can determine the economic events worldwide as a central sustainability instrument, especially also the relationship between the developed and the less developed countries. In this context, the more modest contribution of cannabis, still mainly perceived as a "drug", is also to be seen, whether it is the exploitation of THC (tetrahydrocannabinol) or the THC-purified CBD (cannabidiol).[16]

The medical efficacy of cannabis is so far only inadequately researched; although a wide ethnomedical knowledge[17] is available and the various possibilities of application were already known at the beginning of our century: *"The state of knowledge about cannabis as pharmacotherapy is still expandable."*[18]

Accordingly, the statements of those interested differ considerably:

So the Hemp Association already wrote at the beginning of this century 'How Hemp Helps- Who Hemp Can Help',[19] while Kirsten **Müller-Vahl** and Franjo **Grotenhermen** (2017) inform extensively about the medical use of cannabis after the long-awaited change of the BtMG in the German Medical Journal.

In a very thorough meta-analysis of the relevant scientific literature commissioned by the Federal Ministry of Health in 2019, the editors Eva Hoch, Chris Maria Friemel and Miriam Schneider, authors from the Munich Clinic for Psychiatry and Psychotherapy (LMU), complain under the title 'Cannabis: Potential and Risk' about this overall inadequate research situation, which, among other things, records moderate successes in terms of pain reduction, but is largely still open in the case of mental disorders.[20]

An absolutely still open research situation, which is also confirmed by the position paper by Heino Stöver et al. (2021).[21]

---

[16] For the background: Akzept/Aidshilfe (2019 p. 120–131).

[17] Rätsch (2016).

[18] Johnsen and Maag (2020, p. 4–5).

[19] Geyer (2006, p. 19).

[20] Hoch et al. (2019, p. 27).

[21] Akzept/Aidshilfe (2021, p. 145).

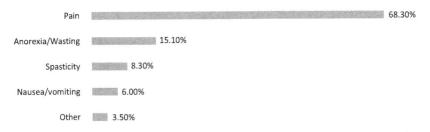

Fig. 4.1 Medical cannabis in Germany. (Statista 2019)

Statistics of Pharmacies

This scientific finding corresponds to the statistics of pharmacies (2018) in Table *"Pharmacy Survey: Should Cannabis be legalized in Germany?"* (Fig. 4.1)[22], according to which cannabis was overwhelmingly bought as a painkiller, while mental disorders were recorded under 'Miscellaneous': Correspondingly shifted data also result for 2020 from the table *"Disease or Symptomatology"*[23], of the BfArM (Fig. 4.2)—of course without private prescriptions,[24]—which is contained in the answer of the federal government to the Left's request of 23.03.2020. The recently published final report of an accompanying survey of the BfArM (2022a) also states: *"More than 75% of the evaluated treatments were due to chronic pain. Other frequently treated symptoms were spasticity (9.6%) and anorexia/ wasting (5.1%)."*

However, Franjo Grotenhermen and Maximilian Plenert[25] criticize in their 'Update on Cannabis as Medicine' in 2020 that *"patients with many other indications are underrepresented"*[26] *here. This would result from the "distribution of diseases for which the Federal Opium Agency has issued an exception permit for the*

---

[22] Brandt (2018).

[23] Bureg/DIE LINKE (2020, p. 4).

[24] Akzept/Aidshilfe (2020, pp. 157–170).

[25] Grotenhermen (2021, pp. 155–162).

[26] Akzept/Aidshilfe (2020, pp. 159 ff.).

Diseases or symptomatology of all complete data sets, as of 6 March 2020	Cases (n=8,872)	Percentage
Pain	6374	approx. 72 %
Spasticity	940	approx. 11 %
Anorexia / Wasting	590	approx. 7 %
Nausea / vomiting	341	approx. 4 %
Depression	259	approx. 3 %
Migraine	181	approx. 2 %
ADHD	111	approx. 1 %
Appetite deficiency /Inappetence	111	approx. 1 %
Intestinal disease, inflammatory, non-infectious	113	approx. 1 %
Bowel disease	55	approx. 1 %
Tic disorder incl. Tourette syndrome	79	< 1 %
Epilepsy	97	approx. 1 %
Restless Legs Syndrome	78	< 1 %
Insomnia / Sleep disorder	74	< 1 %

Fig. 4.2   Disease or Symptomatology. (BfArM 2020)

use of medicinal cannabis flowers from the pharmacy according to § 3 para. 2 of the Narcotics Act". In particular in the case of mental problems and specifically in the diagnosis of ADHD, which made up an above-average proportion in information offers such as the ACM-patient telephone: *"Mental illnesses had a share of 23% among patients with an exception permit. The proportion fell to 5% for cost reimbursements."* The authors attribute this, inter alia, to the *"lack of*[n] *data on the issuance of private prescriptions or private patients"*; but these could only partially cover their needs *"for economic reasons"* due to the lack of cost reimbursement. It can therefore be assumed that *"significantly fewer than 10% of patients—these would be 87,000 people—who need such a therapy, also receive it."*

A situation, for which Heino Stöver et al. (2021) compared to Israel, in which approximately 1% of the population are cannabis patients, while in Germany it is only 0.1%, speak of a *"dramatic under-supply of cannabis-based medications"*[27].

Problems

Since 2017, a new Narcotics Act has been in force, which allows the dispensing of cannabis flowers by pharmacies due to the threat of drug dangers, but regulates it very cautiously[28] and which also provides for their health insurance financing

---

[27] Stöver et al. (2021, p. 5).

[28] Akzept/Aidshilfe (2020, pp. 142–147).

under certain conditions in § 31 of the 5th SGB (Fifth Social Security Code) in a new paragraph 6.[29]

The pharmacies obtain this cannabis from the BfArM for 4.30 EUR per gram,[30] in order to resell it against a narcotic prescription as cannabis flowers for about 22.00 €, while in the Netherlands *"the same packaged cannabis varieties"* can be obtained for 6–7.00 EU/gram.[31] With us, however, the following applies: *"Monthly costs for a therapy are therefore between 300 and 2.200 €. An alternative opioid therapy would be significantly cheaper."*[32] But the three major health insurance companies reject around 1/3 of the approximately 70,000 applications received since 2017.[33] In view of the considerable bureaucratic difficulties,[34] the hesitant behaviour of some doctors and the high pharmacy prices, it is no wonder that patients still supply themselves through self-cultivation today.[35]

These hemp natural products compete with the products manufactured by the pharmaceutical industry, such as dronabinol (THC extract), nabilon, the synthetic derivative of THC available since 1983, or Sativex® (specifically for multiple sclerosis). On a competitive market, which *"measured by prescription numbers* [shows] *"a trend upwards"* [from] 69,000 prescriptions in the 1st quarter of 2020 to 84,000 in the 4th quarter. Cannabis flowers are gaining increasing importance over the year."[36] A future-oriented market, about which the Handelsblatt already wrote in the year of its legalization in 2018: *"In Canada, the market for medical cannabis is booming,"* to predict global demand: *"According to estimates by the market research firm Brightfield Group, the global market for medical cannabis will quadruple to 31.4 billion dollars by 2021."*[37]

---

[29] SGB V § 31.

[30] BfArM (2021).

[31] Grotenhermen (2020, p. 152).

[32] Wohlers (2019).

[33] Telgheder (2021a).

[34] Grotenhermen (2020).

[35] Akzept/Aidshilfe (2021, pp. 148–153).

[36] Johnsen and Maag (2020, p. 4).

[37] Handelsblatt (2018); Aposcope (2022); Schwager (2022).

4.3   SDG 6 and 14: Groundwater and Seas

Although the dwindling freshwater resources in rainfall and groundwater are compared to the amount of salt water in the oceans, both are affected by the same two civilization influences that also threaten our health (SDG 3): Nitrate pollution from the air and from fertilization that enters the sea through rivers, as well as the growing plastic load, whether it be the ubiquitous microplastics or the pollution of the oceans.[38]

As the Nitrate indicator of the sustainability report (Fig. 4.3), which however did not take into account fertilization, but only entries by atmospheric nitrogen,[39] particularly clearly shows, the eutrophication by nitrogen inputs (nitrates) has so far been inadequately reduced: *"Approximately 25% of the groundwater bodies in Germany are in a poor chemical state with regard to the requirements of the Water Framework Directive due to high nitrate values. All transitional and coastal water bodies also fail to reach a good ecological state due to significantly excessive nutrient inputs,"*[40] which is well documented by the graph of the total nitrogen concentration in the North and Baltic Seas (Fig. 4.4)[41].

Phytoremediation
It is not only the case that our soils are becoming more and more acidic, that the set limit values are repeatedly exceeded, but rather these pesticides, nitrates and other toxins get into our soils and groundwater, into our food and animal feed, which we humans in turn take up as products.

A decisive sustainability advantage of hemp is first of all the lower groundwater consumption mentioned below, for example in comparison with cotton cultivation. However, its double ability to use this nitrate (NO_2) and to bind it is decisive. On the one hand, it uses both the atmospheric and the nitrate introduced for its fertilization, such as slurry or chemical nitrates, similarly to the maize plant, directly for its own plant growth. But this nitrate can be bound more permanently in several recycling cycles in the hemp follow-up product on the way of a 'cascade use'. And on the other hand *"it retrieves nitrate from deep soil layers through deep-reaching fine roots"* [...]*where it is no longer accessible to other*

---

[38] Greenpeace (2020).

[39] DNS Weiterentwicklung (2021, p. 336).

[40] DNS Weiterentwicklung (2021, pp. 336–337).

[41] DNS Weiterentwicklung (2021, p. 321).

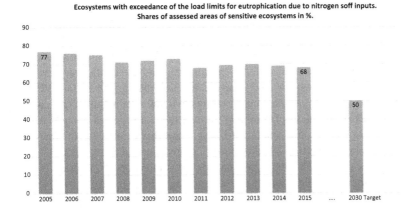

**Ecosystems with exceedance of the load limits for eutrophication due to nitrogen soff inputs.
Shares of assessed areas of sensitive ecosystems in %.**

Fig. 4.3 Ecosystems exceeding the load limits. (Umweltbundesamt 2015)

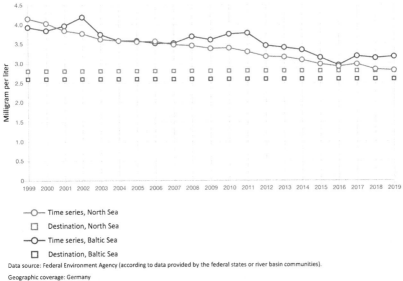

Total nitrogen concentration in the North Sea and Baltic Sea
Concentration in mg/l (moving, discharge-weighted average of the past 5 years).

—O— Time series, North Sea

☐ Destination, North Sea

—O— Time series, Baltic Sea

☐ Destination, Baltic Sea

Data source: Federal Environment Agency (according to data provided by the federal states or river basin communities).

Geographic coverage: Germany

Unit: concentration in mg/l (moving, discharge-weighted average of the past 5 years).

Copyright: © German Federal Statistical Office (Destatis), 2022

Fig. 4.4 Total nitrogen concentration in the North and Baltic Seas. (Umweltbundesamt 2019)

cultivated plants. Therefore good groundwater protection" said the Baden farmers quoted below under the keyword *"Soil improvement by reducing nitrate"*,[42] in order to clean nitrate-contaminated soils in this way.

In this sense, renowned seed banks such as Dinafem[43] and Humboldtseeds[44] rave about how hemp can be used to clean soils contaminated with heavy metals through such phytoremediation. For example, it is said that farmers in Tarent/Italy have succeeded in counteracting a devastating environmental pollution caused by a huge steelworks in the neighbourhood through the cultivation of hemp.[45] And, without giving any specific sources, it is even said that in the case of Chernobyl, the laboratories Phytotech had decided to grow industrial hemp together with some farmers and the Ukrainian Institute for the Cultivation of Fibres in 1998, because it draws strontium and radioactive caesium from the soil, which may rather serve the advertising for the new hemp variety Chernobyl[46] but points to the lack of any relevant research as well as to its necessity.

4.4   SDG 11: Sustainable Cities and Communities

At the beginning of its fourth transformation area (Sustainable Building and Transport Turnaround), the Federal Government writes: *"The construction and building sector is closely linked to the challenges of other transformation areas with its upstream and downstream processes. The requirements for sustainable building include energy efficiency and climate neutrality, the preservation of biodiversity, the conservation and use of renewable raw materials, the reduction of land consumption, the sustainable procurement of products and services including the observance of human rights in the supply chain as well as the safeguarding of the health and comfort of users."*[47]

In fact, construction and transport, in addition to the agriculture discussed below, are the biggest environmental offenders, both as consumers of mate-

---

[42] Pix (2020, p. 11).

[43] Civantos (2017); Dinafem Seeds (2021).

[44] Humboldt Seeds (2017).

[45] Brosius (2016).

[46] Cannaconnection (2021).

[47] DNS Development (2021, p. 56).

rial and ecologically dwindling resources, as well as through their health consequences and through their CO_2-intensive energy consumption.

Hemp could be used here (SDG 9, 11), also as part of the federal funding for
sustainability in construction[48] as a renewable alternative, for example as hemp
lime instead of the extremely CO_2-producing cement production[49] (SDG 13) or as
insulation material, which could permanently reduce energy consumption (SDG
7). In both cases, the use of hemp would both initially permanently store the CO_2
absorbed during growth (SDG 13), and later serve as the basis for 'urban mining'
as a model of a circular economy (SDG 12), instead of polluting the environment
as enormous amounts of construction waste. A circular economy, as is already
known today in the field of paper and plastic, and which could be significantly
expanded with the help of hemp in the construction industry.

On the one hand, the constantly lamented new residential construction with its
need for 'alternative' building materials is in question, and on the other hand, the
large mass of existing buildings with their renovation needs, especially in the area
of thermal insulation.

New Buildings
With regard to new construction, the independent *empirica institute* estimates the
need for new housing, without catch-up demand for 2020/21, at just over 270,000
apartments with a slightly declining trend overall,[50] while the Federal Government provides for 400,000 apartments annually.[51] Here, hemp stones[52] produced
with the help of hemp and hemp lime wall, also called hemp concrete, come into
consideration, which *"consists of hemp shives mixed with lime as a binder and
water, which is preferably used for non-load-bearing walls,"*[53] in order to then
possibly use it directly in the currently tested spray-pressure process for house
construction in the future.[54] If one compares this with the conventional building materials, then one ton of steel causes 1.46 t CO_2, one ton of steel concrete

---

[48] DNS Development (2021, p. 61); BMI (2021).

[49] Römer (2019).

[50] Braun (2020, p. 4).

[51] Bureg (2022a).

[52] Schönthaler (2022).

[53] Grimm (2020).

[54] Römer (2021).

198 kg CO_2 and cement 587 kg CO_2 per ton.[55] While, regardless of the energy costs for transport and assembly, hemp stone is up to 90% CO_2 negative due to its CO_2 binding.[56]

This hemp lime has, apart from its relatively low load-bearing capacity, only *"actually only building physics advantages. Its thermal insulation corresponds to that of modern masonry bricks with insulation filling. External walls made of hemp lime therefore do not require an additional insulation layer. Hemp lime also retains its good insulation value even in damp condition. The lime binder makes the material alkaline and thus antibacterial and mould-resistant. Since the plant components are completely enveloped by the mineral binder, there is also no destruction by vermin. In addition, the composite material is not combustible."*[57]

Renovation

As a renovation basis, an annual 1 million dwellings[58] and non-residential buildings are taken into account, for which, however, there are no statistical figures, which is why the Federal Institute for Building, Urban and Spatial Research states in 2016: *"Some assumptions assume 2.0 million buildings, other studies up to 3.3 million existing non-residential buildings in Germany. Especially the number of industrial buildings is very uncertain."*[59]

So far, the corresponding renovation activity in the building envelope has been around 1% annually, without making any statement about the renovation depth[60], while the Federal Environment Agency 2019 in its background report 'Living and Renovating Empirical Building Data' since 2002 (without further specification of the year) for Germany as a renovation status of residential buildings: Unrenovated: 35.9%, partially renovated: 51.4%, fully renovated: 4.3% and new construction: 8.4%,[61] and already proposed a renovation rate of 2% in 2014.[62]

---

[55] WWF (2019, p. 7).

[56] Schönthaler (2022).

[57] Grimm (2020).

[58] BMWI (2014).

[59] BBSR (2016).

[60] BMWI (2014 p. 5).

[61] Federal Environment Agency (2019, p. 76).

[62] Federal Environment Agency (2014, p. 5).

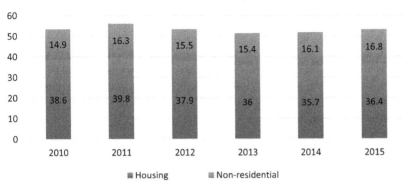

Fig. 4.5 Development of energy measures in residential and non-residential construction in billion euros (BBSR 2016, p. 6). (Own representation)

The market-based order of magnitude in the billions here illustrates the diagram (Fig. 4.5) for the years 2010 to 2015.[63]

Two aspects are relevant here for the use of hemp, on the one hand the energetic renovation of existing buildings by insulation and on the other hand the reuse of the building material used in the sense of urban mining.

Insulation
Insulation with hemp material can be used in particular for old building renovations. The Federal Government declared in its answer of 21.12.20 to a request from the FDP, that only 7% of the raw materials used for insulation come from renewable raw materials. The total market corresponds to 28.4 Mio. m^3, of which only 2.0 million m^3 from renewable raw materials and 100,000 m^3 of hemp insulation, i.e. 0.35% of the entire market, although hemp boards require only a tenth of the energy compared to mineral wool.[64] It would be positive to note that there is currently a change of mindset among the builders, so that the demand for natural insulation materials would experience a significant increase *of 3.1% per year.*

---

[63] BBSR (2016, p. 6).
[64] Klöckner (2018, p. 11).

The demand is therefore increasing.[65] *But even the insulation of facades* [...]*is not undisputed, since the insulation materials are difficult to recycle and can also lead to mold, because the houses then "sweat"*[66], which should not apply to the hemp boards in particular.[67]

However, prices have risen sharply in recent years due to high demand and low stocks. Thus, depending on the type and thickness, the square meter cost around 4 to 30 € in 2016,[68] while it currently costs between 34–55 € per square meter as a 'luxury insulation material',[69] which is why it would be particularly sensible to include this alternative in the focus 'insulation' of the 'Renewable Raw Materials' funding program.[70]

Urban Mining

Already today, the construction industry in Germany is responsible for the extraction of 517 million tonnes of mineral raw materials and 14% of total greenhouse gas emissions. The waste generated by the construction industry accounts for 52% of German waste;[71] *"most building materials are* [however] *hardly reusable."*[72] While building materials made of hemp can be recycled with lime and water according to the 'Cradle to Cradle' principle or can be composted.[73] Such a 'cascade use'[74] would correspond to the sense of a consistent circular economy, in which this material could be reused as 100% recycled hempstone without producing any further waste.

---

[65] Bureg/FDP (2020, p. 2).

[66] Götze (2021).

[67] Grimm (2020).

[68] Energy experts (2016).

[69] Stiftung Warentest (2021b).

[70] FNR (2022).

[71] DNS Development (2021, p. 55 ff.).

[72] Götze (2021).

[73] Schönthaler (o. J.).

[74] DNS Development (2021, p. 286).

4.5   SDG 12: Sustainable Consumption and Production

The wide field of this goal 12 becomes clear when, on the one hand, the globalizing introductory remark of the Federal Republic: *"SDG 12 aims at the necessary change of our lifestyles and our economic system. Sustainable consumption and sustainable production require us to consume and produce today in such a way that the satisfaction of the legitimate needs of current and future generations is not endangered, taking into account the limits of sustainability of the earth and universal human rights as well as the other sustainability goals"*, on the other hand, is compared with one of its indicators (No. 12.3a) 'Share of paper with Blue Angel in total paper consumption of the direct federal administration'.[75]

From the short and long hemp fibers, together with flax or even sheep's wool,[76] further material alternatives with sustainable properties can be produced. Traditionally preferably for paper production and in the field of textiles, but nowadays also increasingly as a plastic substitute and in the automotive sector because *"Hemp fibers are just as stable as glass fibers, around a third lighter, tear-resistant and can be disposed of or recycled in an environmentally friendly manner,"*[77] while the shives provide the cellulose necessary for plastic.

Hemp Paper[78]

The classic hemp product was and is hemp paper with currently about 70–80% of the hemp fiber market, which today mainly flows into the *"production of banknotes, cigarette paper and hygiene products"*. Compared to the use of timber, hemp paper *"lasts longer and is tear-resistant and can also be used in a wet state"*. The use of chemicals, due to the very low lignin content, is also lower than in the current paper production. As an annual plant, hemp delivers 4–5 times as much paper on the same area as a forest and can also be recycled more often than paper from wood.

---

[75] DNS Development (2021, p. 286, 301).

[76] Wieland and Bockisch (2003, p. 2).

[77] Hempopedia (o. J.a).

[78] Hempopedia (o. J.b).

Textile Sector, Clothing, Cottonized Fibers

For millennia, hemp has been used to make clothing, ropes and sails, which today could be used in this 'raw' form for geotextiles in civil engineering and water engineering or for hemp felt carpets, for example in the exhibition sector, due to its compostability, instead of the subsidized cotton of the farmers or industrially produced synthetic fibers such as polyester. But *"today one of the most interesting product lines for German hemp"* [...] *"lies in the textile sector."*[79] In particular for the modified *"cottonized hemp fibers"* known since the 20s of the last century: *"This refers to a hemp fiber that has been refined using modern chemical-physical processes and is so similar to the cotton fiber in its technical properties that it can also be processed on highly productive cotton spinning machines."*[80]

So you can buy ladies' and men's fashion at *Hanfare*[81]. H&M also uses organic hemp, among other things, in its *Conscious Exclusive Collection*.[82] Even fashion houses are now rediscovering hemp. The *"Ralph Lauren fashion house has used hemp-silk charmeuse made by EnviroTextiles to make intensive use of it in its clothing"* and has made evening dresses and even a military jacket using various hemp mixtures.[83]

Again, this fabric group also has to be considered from the double sustainability perspective: On the one hand, the fabric made of hemp is not only more resistant, antibacterial, resistant to mildew, mold and UV radiation[84], but is also completely recyclable, which is far superior to the other fabrics used to make clothing.

On the other hand, the cultivation, production and processing are more resource-friendly for the environment, as not only less water and chemicals are consumed. According to the *Stockholm Environmental Institute*, cotton requires 7.58 to 9758 liters of water per kilo, while hemp is 75% lower, only 2401 to 3401 liters of water per kilo.[85] A finding that the Leibniz Institute for Agricultural Engineering and Bioeconomy in Potsdam was able to confirm in a field trial in 2017 and 2018: The *"cannabis varieties specifically bred for growth conditions in*

---

[79] Bócsa and Karus (1997, pp. 150 ff.).

[80] Nette-group (2015).

[81] Hanfare (o. J.).

[82] Riehl (2019).

[83] Sensi Seeds (2020).

[84] EIHA (2020, p. 26).

[85] Barrett (2020).

Europe are suitable for cultivation on rather dry sites. According to the research-ers, the water productivity of the hemp varieties examined was about six times higher than that of cotton. This [...]therefore has to be grown on moist sites or irrigated."[86]

Plastic

The plastic consumption based on petroleum is today one of the decisive environ-mental problems, not only because of the associated CO_2 production, but above all because of the associated waste production, which particularly burdens the oceans.

A special problem are the microplastic waste,[87] which, with their effects insuf-ficiently examined so far, also enter our food through field crops and fish.

That is why many disposable plastic products have been banned in the EU since July 3, 2021. In addition, since January 1, 2021, there has been an EU-wide export ban on heavily recyclable plastic waste that is mixed or contaminated.[88]

Accordingly, even *"a group of global companies from the plastic and con-sumer goods value chain,"* [...] *"has founded a new non-profit organization to drive solutions to eliminate plastic waste in the environment, especially in the sea. The Alliance to End Plastic Waste (AEPW) currently consists of almost 30 member companies and has committed more than one billion US dollars to develop and implement solutions to minimize and manage plastic waste. The World Business Council for Sustainable Development (WBCSD) is a strategic partner of the Alliance."*[89]

The high CO_2 emissions associated with plastic production are particularly burdensome, as an international *"network of climate protection actors"* [there named] *"has found out," "because from production to disposal, climate-damag-ing CO_2 enters the atmosphere. In 2019 alone, this results in 850 million tonnes of greenhouse gases, comparable to the emissions of 136 coal-fired power plants in the same period."*[90]

---

[86] Podbregar (2020).

[87] Federal Environment Agency (2015).

[88] Bureg (2021).

[89] AEPW (2019); Gassmann (2020).

[90] Greenpeace (2019).

The well-recyclable natural product hemp could be used here in a particularly meaningful way: It provides with 2 to 3 t/ha[91] *"valuable cellulose—a component of cell walls. Not only paper can be made from it, but also plastics: for example cellophane (cell glass), the cotton substitute viscose and celluloid (cell horn)."*[92] For example, the US company Sana Packaging *"designs and develops different, sustainable and compliant packaging solutions for the cannabis industry using 100% plant-based hemp plastic."*[93] But here, too, there are still problems to be solved by research and politics. For example, the often used biocomposite products,[94] such as those coated, are difficult to separate and no longer compostable: *"So one cannot really speak of a real alternative, unless there is a drastic change in the disposal infrastructure in Germany. In addition, the ecological balance sheet of bio-plastics would have to be improved, as the production still has to struggle with a high energy consumption."*[95]

Car Production

Among hemp enthusiasts, reference is often made in this context to the *hemp car,* presented by Henry Ford in 1941, whose body was ten times stronger than sheet metal, which is why it weighed *"900 kg, about 450 kg less than a car with a metal body"*,[96] and which was driven with biofuel from hemp oil: *"The cars panels were moulded under hydraulic pressure of 1500-pound psi from a recipe that used 70 percent of cellulose fibres from wheat straw, hemp and sisal plus 30 percent resin binder,"*[97] which, however, resulted in only a 10% hemp content[98]. Recently, Bruce Dietzen developed his own CO_2 negative sports vehicle based on this model in 2017, called RENEW: *"to build a car body that's made up by 100 pounds of cannabis, all covered in an extremely hard resin."*[99] *"The Renew is a true sports car that can be configured to 80 horsepower or a 525 horsepower dragster with a Flyin' Miata drivetrain. Weighing just 2500 lbs, the turbo package*

---

[91] Bòcsa and Karus (1997, p. 32).

[92] World of Wonders (2021).

[93] Greenvision (2019).

[94] Omar et al. (2012).

[95] World of Wonders (2021).

[96] Plumb (1941).

[97] Dutta (2018).

[98] PottsAntiques (2010).

[99] Jacobs (2017).

gives a weight-to-horsepower ratio comparable to a Porsche 911 Cabriole gushes Ganjapreneur."[100] But Porsche also used an *"organic fibre mix"* for its new 718 Cayman GT4 Clubsport for car doors and rear wings, *which are primarily sourced from agricultural by-products such as flax or hemp fibres and feature similar properties to carbon fibre in terms of weight and stiffness."*[101]

Already today, *"Faurecia, one of the leading technology companies in the automotive industry,"* sells NAFILean™ (Natural Fibres for Lean Injected Design) in the USA for *"structural vehicle parts such as dashboards, door panels and centre consoles, which are then covered with leather or fabric upholstery. The injection moulding material based on natural fibres combines natural hemp fibres with polypropylene resin and enables the production of complex shapes and structures while simultaneously reducing weight."*[102] Parts that are installed, for example, by PSA(Peugeot Citroën DS Opel), FCA(Fiat Chrysler), JLR (Jaguar Land Rover) and in the 'Megane' by RSA (Renault Group, including Nissan and Mitsubishi).[103] And *Hempopedia* reports that BMW, Ford and Daimler also use such fibre composites including hemp for door and boot linings, dashboards and armrests: *"A statistic shows that natural fibres (flax, sisal, jute and also hemp) are increasingly being used in the automotive industry. Currently, about 19,000 t are used per year in German car production, compared to only about 9 t in 1999."*[104]

In addition to the typical positive hemp properties: recyclable natural materials, low water and pesticide consumption, hemp could specifically in body construction *"offer a high passive safety"*, *since the parts break off dull and do not form sharp edges"*:[105] *"The stable components are particularly sound-absorbing and convince in terms of crash properties."*[106] Even more important would be a considerable reduction in weight in road traffic due to its lighter weight, which could particularly benefit battery-laden e-mobility[107] shift the distance/energy consumption ratio significantly in its favor—less weight, therefore higher range

---

[100] Abbott (2020).

[101] Porsche (2019).

[102] Faurecia (2018).

[103] Demortain (2018).

[104] Hempopedia (o. J.a).

[105] Frahm (2013).

[106] Klöckner (2018, p. 13).

[107] Bureg (2020).

or less energy consumption—(SDG 13). That is why corresponding research funds should be used here for its further development, for example in the context of the 'Lightweight Initiative' of the German government[108] (SDG 8, 12). So *faurecia* writes on 3.14 2018: "*[T]he first commercial success was achieved in 2013 in door panels for the Peugeot 308. The 1.2 kgof the material delivered a 25% weight reduction and a 25% environmental impact reduction*" and *calculates in an interesting calculation*" [A] *reduction of 40,000 tons of* CO_2 *emissions and the ability to drive an additional 325 million kilometers with the same quantity of fuel.*"[109]

These are the "small" differences that, in principle and in the future, in view of the number of cars in Germany of around 48.25 million as of January 1, 2021 and worldwide of 1,197.71 million (2020)[110] would add up to considerable CO_2 savings if hemp fibers were used more extensively in car construction instead of other fiber plants, such as the Abaca banana fibers from the Philippines[111] favored by Mercedes.

4.6   SDG 13: Measures for Climate Protection

The production of CO_2 is considered the main cause of the worsening climate crisis, its reduction is a main concern of the Climate Protection Act of August 31, 2021, according to which emissions should be reduced by 65% compared to 1990 by 2030 and climate neutrality should be achieved by 2045.[112] For this purpose, *"the annual emissions in agriculture should be reduced by 14 million tonnes of* CO_2*-equivalents compared to 2014."* For which, although the *"expansion of organic farming"*, *"soil retention and build-up in arable land"* *"moorland protection"* and *"preservation and sustainable management of forests and wood use"* are listed, cannabis is not listed.[113] Cannabis/hemp could, however, contribute to this worldwide in three ways: directly as a plant in analogy to the

[108] BMWI (2021).

[109] Demortain (2018).

[110] Statista (2021); Umweltbundesamt (2020).

[111] Knauer (2005).

[112] Climate Protection Act (2021).

[113] BMEL (2020).

call to plant trees as CO_2 storage; indirectly in comparison to its possible alternatives, and indirectly through a reduction in transport costs.

State of Knowledge: Government; Research

The Federal Government said on July 4, 2019 in its answer to the Left's inquiry in point 13: *"How does the Federal Government assess the contribution of hemp cultivation in agriculture to climate protection, and what conclusions does it draw from this, in particular with regard to [...]CO_2 binding":"So far there are no measurements of greenhouse gas balances of hemp. Due to the lower nitrogen requirements and lower costs for cultivation, it is to be expected that the greenhouse gas emissions per hectare are lower than with other crops (e.g. wheat, rape, maize). It is also assumed that the binding of carbon in the soils is significantly influenced by the type of use of hemp."* [...]*In general, hemp has an intensive and deep rooting. This can promote the binding of carbon in the organic soil substance.*[114]

In this sense, the European Industrial Hemp Association (EIHA) claims in its Hemp Manifesto 2020[115]: *"One hectare of hemp can absorb up to 13.4 t CO_2 and is thus as efficient as one hectare of rainforest."* While James Vosper in 'The Role of Industrial Hemp in Carbon Farming' even calculates that *"One hectare of industrial hemp can absorb 22 tonnes of CO_2 per hectare."*[116] This can be compared to the result of the Vienna University expert opinion by Andreas Richter et al. (2011) for the calculation of the CO_2 savings through a replantable newly planted rainforest, in which they calculate an *"average annual carbon binding over 60 years"* [of] ~ *10 t CO_2 per hectare [...]and year in this reforestation project"*[117]. As complex as these calculations are, not only soil, varieties and fertilization come into play here, but also the hemp mass obtained per hectare, the calculation of rooting, humus formation, etc., so the finding of a significant CO_2 improvement given by the Federal Government should be correct in the end result.

This is also shown in the equally complex comparison of different arable crops—without hemp—by Prof. Dr. Schönberger and Pfeffer (2020):

---

[114] Bureg/DIE LINKE (2019, p. 10).

[115] EIHA (2020, p. 2).

[116] Vosper (2020).

[117] University of Vienna (2011).

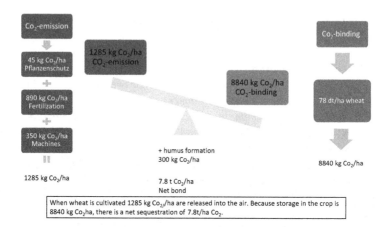

Fig. 4.6 CO_2 balance as an example of wheat cultivation. (Based on Schönberger and Pfeffer 2020 p. 65)

'Agriculture: CO_2 sinners or saviors?',[118] which I can also use here as an example of a possible hemp calculation using the graph (Fig. 4.6) of the CO_2 binding in wheat cultivation. Here, in the end, only a 7.8 t/ha net binding was obtained, namely, under additional calculation of the CO_2 expenditure for plant protection, fertilization and machine operation, which, of course, would also have to be taken into account in a reduced form for such a 'hemp comparison'.

The research tasks still necessary here, which have so far been hardly realized, can be demonstrated by the example of the Bavarian research project by Veronika Schöberl et al. (2019): 'Hemp for material use: status and developments', which found in a complex field project that depending on the variety and level of nitrate fertilization a dual yield (seed + fiber) of 8 to 11 t/ha can be achieved.[119]

Primary CO_2 Net Gain

The most important contribution to sustainability of hemp in the face of climate crisis is its primary 'positive CO_2 balance', comparable to afforestation, which both takes CO_2 from the air for plant growth and potentially stores it permanently through 'cascade use', as well as anchors CO_2 in the soil through humus

---

[118] Schönberger and Pfeffer (2020).

[119] Schöberl et al. (2019, p. 14).

formation. This plant-based CO_2 removal (Carbon Dioxode Removal—CDR)is recommended by the Intergovernmental Panel on Climate Change (IPPC) as a biological storage (BECCS) more effective than the *"very energy-intensive"* and *"not yet available on a large scale"* *"Direct extraction of CO_2 from the atmosphere with subsequent storage (DACCS =Direct Air Capture and CCS)."*[120]

Secondary CO_2 Gain

This direct or primary CO_2 net gain of the plant is significantly, although still largely lacking in an exact calculation, increased in the course of its previously described utilization. As mentioned in the previous sections, firstly through its relatively permanent binding in construction, such as hemp stone and hemp lime or as a recyclable insulation material, which both replaces energy-intensive alternatives and saves CO_2-rich heat regulation, but also as food or in horse bedding, which then also promotes humus formation.[121] Its use in car manufacturing could, with reduced weight, not only enable the desired higher ranges with E vehicles, but also significantly reduce the CO_2 sins of fuel consumption. In the processing of textiles, for example in comparison to the production effort for cotton and wool, and especially as an alternative to plastic production, one could save tons of CO_2, regardless of its other ecological advantages, from resource consumption to marine pollution.

Tertiary CO_2 Gain in Transportation

Above all, however, and mostly unnoticed, the savings possible in transportation support the goal of CO_2 reduction. And again in double form. On the one hand, sufficiently promoted domestic cannabis production should reduce the considerable costs of imports today, which currently exceed domestic production due to its excessive costs due to lack of mass. And on the other hand, such a small-scale, peasant economy, analogous to the situation mentioned below in Austria,[122] would have a CO_2-reducing effect on traffic instead of the current, large-scale industrial truck queues.

---

[120] DNS Development (2021, p. 305).

[121] Bòcsa and Karus (1997, p. 157).

[122] WVCA (o. J.).

Also with 'Medical Hemp'

These various primary to tertiary CO_2 reductions generally apply in the same way to both industrial hemp and medical hemp, primarily in cultivation, and secondarily both in comparison to chemically-industrial pharmaceutical production and on the transport level. Especially since medical cannabis is still imported to a large extent today and has to be handled bureaucratically by the Cannabis Agency and the company Cansativa. A start-up that regulates the procurement, supply, storage and transport of medical cannabis for all pharmacies in Germany, for which only two locations in Germany have been certified so far.[123] Especially since this transport requires considerable energy expenditure, as the example of the transport of medical cannabis worth 5 million euros from the Medical Cultivation Laboratory in Portugal[124] by the company Tilray showed in 2019:[125] since it can only be transported between 15 and 25 degrees, extra air-conditioned cabins are required for this.[126]

4.7   SDG 15: Life on Land

If one wants to follow the causal-logical development path of the hemp effect, it is advisable to look at agriculture in particular, which sets its sustainability profit in motion above all through the expansion of industrial hemp.

The Federal Government emphasizes this goal 15: *"Intact ecosystems are an indispensable basis for human existence and sustainable development. They are the basis for ensuring a diverse diet, provide clean air and clean drinking water resources and provide important raw materials,"* in order to connect this goal 15 *"as a cross-cutting issue"* [with] *"many other SDGs"*: SDG 2 (food security), SDG 6 (water), SDG 11 (sustainable urban development), SDG 12 (sustainable consumption and production patterns), SDG 13 (combating *climate change) and SDG 14 (seas),"*[127] while the FAO (Food and Agriculture Organization of the UN)[128] speaks of 'sustainable livelihoods and food systems' to capture the complexity of this goal.

---

[123] Endris (2020).

[124] Winkler (2019).

[125] Tilray (o. J.).

[126] Winkler (2019).

[127] DNS Weiterentwicklung (2021, p. 326).

[128] FAO (2020).

Requirements: Cultivation of Industrial Hemp

In Germany, an estimated 16.6 million hectares (ha) were used for agricultural purposes in 2020, including 2.6 million ha for renewable raw materials (NawaRo), i.e. for *"energy and industrial plants,"* which is 15.7% of this area. This includes 4650 ha for fibre plants, *"essentially hemp,"*[129] which would be 0.18% of these NawaRo and only about 0.03% of the agricultural area, while maize consumed about 6% of this area. According to information from the Federal Government of 21.12.20, industrial hemp has been cultivated again in Germany since 1996, with a total of 5362 ha in 2020. Pioneers are Lower Saxony with 1,105.0 ha, accounting for around 21% of German cultivation, closely followed by Mecklenburg-Vorpommern with 16% and one of the laggards, Saxony-Anhalt, with only 4.2%.[130]

However, the cultivation of industrial hemp in Germany is currently surrounded by rules and bureaucratic regulations, for which the brochure of the Federal Institute for Agriculture and Nutrition from January 2021 provides a frightening overview.[131] These range from EU regulations to the BtMG, which in Annex 1 to § 1 para. 1 lists most of the EU-approved varieties and only allows farmers with a retirement insurance, in BtMG § 19 para. 3 for monitoring the Federal Institute for Agriculture and Nutrition used and in § 24a BtMG the notification obligations regulated, up to the brochure "Culture Guide" (2017) of the Competence Centre for Organic Farming Lower Saxony GmbH.[132]

Problems of Agriculture

As problems of agriculture, which one could not solve but at least mitigate with the cultivation of hemp, in addition to an insufficient development of organic agriculture (Fig. 4.7):[133] Drought as a result of climate change,[134] the unequal, hectare-based distribution of subsidies, through which large farms with their monocultures[135] are favored and small farms are disadvantaged,[136] the resulting

---

[129] BMEL (2021).

[130] Bureg/FDP (2020, p. 1–2).

[131] BLE (2022); BLE Nutzhanf (o. J.).

[132] Rolfsmeyer (2017).

[133] DNS Development (2021, p. 145).

[134] Kuebler (2020).

[135] Rösemeier-Buhmann (2021).

[136] Proplanta (2021).

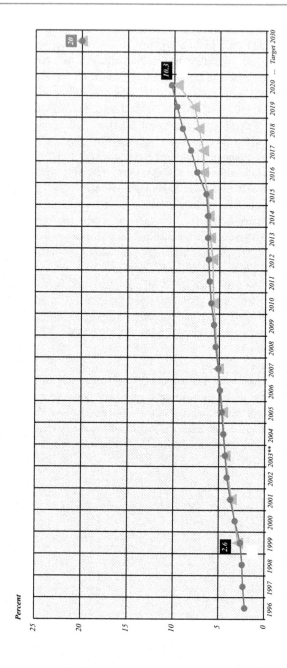

Fig. 4.7 Agricultural areas under organic management. (Federal Statistical Office)

loss of biodiversity of plant and insect species[137] and above all the excessive use of insecticides, pesticides and fertilizers (PSM) caused by it, which the even the European Court of Justice (ECJ) had criticized in its judgment of 21.6.2018.[138]

Hemp Economy

The cultivation of hemp could initially set a symbolic, but significantly expandable accent in reducing these problems. Climatically, this would mainly be due to the reduced CO_2 consumption mentioned. Economically, hemp cultivation, with sufficient subsidization, would initially help mainly small farms, which are currently neither sufficiently supported by the EU program nor at the state level. For example, Baden farmers complained in the paper *'Hemp in the Sign of the Climate Crisis, Sustainable Agriculture, and the Raw Material Turnaround'* (January 2020), which I also refer to in the following, that legumes received 700.00 € per ha[139], while they themselves went empty-handed: *"With a subsidy of 700.00 €/ ha"* [...] *"this would reduce the price to approx. 0.50 €/kg and we would have the possibility to produce local hemp seed oil for approx. 2.00 €/Ltr."*[140] The Federal Government is also considering in its response to the aforementioned Left Party inquiry (4.7.2019): *"Overall, under the assumptions made for hemp cultivation DAKfL [Direct and labor-free performance] from 100 to 250 euros/ha, which are significantly below wheat cultivation (340 euros/ha). From the results it can be concluded that hemp cultivation is currently a niche culture that is economically attractive mainly for organic farming. Due to the low market volume, any funding programs are not considered to be sensible [!]. In such a scenario, due to an induced rapid expansion of supply through support, there could be considerable price reductions in the current high-price segments of hemp seed and CBD."*[141]

For example, the 'Cannabis Austria Economic Association' (WCA) founded in 2018 would like to promote this peasant economic factor, which can point out on its website that in Austria already 300 registered companies with *"1500 qualified[n] and well-paid[n] jobs[n] in a future-relevant industry [...]a turnover of approx. 250 million euros"*[142] could generate.

---

[137] Böhning-Gaese (2021); Walch-Nasseri (2022).
[138] Federal Environment Agency (2018).
[139] Pix (2020, p. 6).
[140] MRL Baden-Württemberg (2020).
[141] Bureg/DIE LINKE (2019, p. 9).
[142] WVCA (o. J.).

Hemp Ecology

The aforementioned farmers, who have been growing hemp in Gottenheim near Freiburg in the Grünbachgruppe[143] water protection area, the first German hemp cultivation area of recent times, for 25 years (since 1996), emphasize the otherwise little-known agricultural advantages from their experience. In addition to the low use of pesticides, there is no economic need for the use of chemical fertilizers and, in addition to the omission of herbicides, due to its 'effective weed suppression', and because hemp *"is very sensitive to most herbicides"*, these farmers emphasize the *"less drying of the soils, thus less watering necessary"*; an advantage that is particularly clear compared to the cultivation of cotton.

The cultivation of hemp also serves the—so topical today—erosion protection, which prevents *"washing out in heavy rain"*. The farmers particularly emphasize its *"high pre-crop value, as it opens and aerates the soil in deep layers through its deep-reaching roots"*, so that one can when integrating *"25% hemp into the crop rotation, e.g. grain after hemp"* obtain *"10–15% more yield"*, especially since it is sown in April as a pre- and intercrop.[144] Why a practical test by the Bergische University of Wuppertal could prove that winter hemp *"in direct comparison with summer hemp"* also provides *"additional yield"* [without] *"competing with other main crops"*[145].

Ecologically, hemp promotes biodiversity. Thus, the Federal Government in its aforementioned answer: *"Hemp is an old crop in Europe. Its use in crop rotations enriches the genetic diversity in the cultivation system. According to a rough evaluation system, set upby Montfort and Small, of the "biodiversity friendliness" of globally cultivated main crops, based on 25 parameters for assessing the environmental impact of the crops, hemp oil plants ranks 3rd and hemp fiber 5th in a comparison of 23 crops."*[146] In particular, hemp, the above-mentioned report of the farmers continues: *"a large variety of insects and thus also serves the bird world"* since *"the hemp crop is harvested relatively late"* it also offers *"an optimal retreat for wild animals and insects."* This is also true for the endangered bees due to the particularly high pollen production of the hemp plant[147], as a research project of Cornell University, New York, could show: *"Because of its*

---

[143] Pix (2020).

[144] Bòcsa and Karus (1997, pp. 131 ff.).

[145] FNR (o. J.); Rinklebe (2019).

[146] Bureg/DIE LINKE (2019, p. 9).

[147] Bòcsa and Karus (1997, p. 37).

temporally unique flowering phenology, hemp has the potential to provide a critical nutritional resource to a diverse community of bees during a period of floral scarcity and thereby may help to sustain agroecosystem-wide pollination services for other crops in the landscape."[148]

Consequences for Sustainability

In the context of SDG 15, the cultivation of hemp therefore offers four sustainability approaches: First, as mentioned above under SDG 2 (nutrition), as a healthy food for humans and animals. Secondly, as explained under SDG 6 and 14, such cultivation reduces the nitrate load on the soil and thus on the groundwater (SDG 6) and thus once again our health (SDG 3).

Then the hemp cultivation could reduce the deforestation in and from the forest through its cellulose offer, nationally and internationally.

A third essential sustainability advantage results, especially in comparison to other field crops, from its ecological balance, to which the mentioned FAO based on a meta-analysis and two field studies, which however did not concern cannabis, states that a *"robust scientific evidence for the strengthening of resilience through agroecology"* speaks, namely *"specifically by strengthening ecological principles, in particular biodiversity, diversity in general and healthy soils."*[149] That is why the Federal Government has set itself the goal in its *"Future Strategy Organic Farming"* [...]*to increase the proportion of areas under organic management to 20% by 2030.*[150] This would also correspond to the UN Biodiversity Convention concluded in Rio de Janeiro in 1992, which had been signed by 196 contracting parties by February 2021.[151]

This initially concerns soil improvement through an extended crop rotation, the low fertilizer and pesticide doses as well as the chance of a nitrate improvement, which in turn can have an effect on the groundwater as well as on the seas (SDG 6, 14, 15). Last but not least, hemp cultivation also promotes bee and other insect protection in the area of biodiversity, for which the Federal Cabinet decided on the 'Action Programme for Insect Protection'[152] in September 2019, and for which the 'Plant Protection Application Ordinance' entered into force on

---

[148] Flicker et al. (2020); Orlowicz (2020).

[149] FAO (2020).

[150] DNS Development (2021, p. 65).

[151] BFN (o. J.).

[152] Bureg (2019).

8 September 2021,[153] which not only restricts or prohibits glyphosate, but also provides for certain protected areas.

Finally, economically speaking, the promotion of hemp cultivation could promote small and medium-sized family farming, among other things, because "the high transport costs for hemp straw [...] *force a regional processing*" in the first processing stage,[154] as the examples from Freiburg and Austria show:[155] Which is why the German Sustainability Report rightly states: "Due to its principles (e.g. *circular economy*, land-based and particularly animal-friendly husbandry), the conversion to organic farming offers small and medium-sized family businesses a development perspective for the future." This could increase the productivity and direct income from agricultural work, for example by eliminating costly additives, through improved crop rotation and winter hemp cultivation, as well as by reducing demand from the construction and food sectors. Within the framework of the "*Improvement of Agricultural Structures and Coastal Protection*" (GAK) program, the "most important national funding instrument to support agriculture and forestry, development of rural areas [...]", one could promote rural economies and thus reduce the significant inequality between rural and urban areas (SDG 10).[156] This would be even more true if new industries were to be created in rural areas and the long transport routes, both nationally and in imports, were to be reduced (SDG 8, 9, 11,15).

Overall advantages, which is probably why the new Minister for Agriculture, Cem Özdemir, announced a large-scale hemp cultivation in Germany at the beginning of his term to finally end the "*madness of the cannabis ban*": "*Many farmers are ready to start growing hemp.*"[157]

4.8   SDG 16: Peace, Justice and Strong Institutions

As much as Goal 16 and Goal 17 (Partnerships for the Goals) have an eye on the international level, which has so far been regulated in the cannabis area under the drug criminalization perspective through the three international drug conventions

---

[153] BMUV(2019); Bureg (2022).

[154] Bòcsa and Karus (1997, p. 149).

[155] Pix, (2020); WVCA (no date).

[156] DNS Weiterentwicklung (2021, pp. 147, 255); BMEL (2022a).

[157] Business Insider Deutschland (2021).

of 1961, 1971 and 1988,[158] the Federal Government then demonstrates with its indicator No. 16.1 that the decreasing number of crimes per 100,000 of the population is making the security feeling of the German population grow: *"A safe environment in which citizens can live without fear of arbitrariness and crime is an essential prerequisite for sustainable development. Therefore, the number of recorded crimes per 100,000 inhabitants should be reduced to less than 6500 by 2030."*[159] In the area of cannabis consumption offenses, however, *"225,120 cases—three percent more than in the previous year* [pursued by the police],"[160] were recorded in 2019, while in 2020 a total of 40,331 people, overwhelmingly cannabis consumers, were convicted of 'unauthorized possession of narcotics' (BtMG § 29 para. 1, sentence 1 no. 3).[161]

However, a legalization of this cannabis consumption, which is not to be discussed in more detail in this work, would 'sustainably' not only end the suffering and discrimination of these consumers, but, according to a calculation by Justus Haucap, Professor of Economics, from 2018, would also release considerable public funds, among other things for the research of the sustainable contribution of hemp, because: *"The 1.1 billion euros in saved costs in the enforcement of law thus only reflect a conservative lower limit. Overall, anamount of 2.66 billion.Euros can be generated through legalization, on the one hand through tax revenues and on the other hand through saved expenses."*[162] An amount that Haucap (2021) now puts at 4.7 billion in a new edition of his calculation.[163]

4.9 SDG 4: High-Quality Education and Research

The Federal Government understands the goal 4 as well as the UN in their two reports, relatively narrowly defined, essentially the school and university education. But already the more extensively quoted South Baden hemp farmers planned for August 2020 'Baden Nutzhanftage' ('days of industrial hemp'), which unfortunately fell due to Corona, on which: *"not only renowned speakers will report*

---

[158] UNODC (o. J.)

[159] DNS Weiterentwicklung (2021, p. 337).

[160] Police (2020).

[161] Statista (2020a, p. 48).

[162] Haucap (2018).

[163] Haucap and Knoke (2021).

on existing initiatives, but there will also be various practical seminars in which the population gets to know hemp as a regional, sustainable building material (hemp lime, insulation wool), experiences the potential of the strongest native natural fiber and learns about the great importance of high-quality hemp seed oil and hemp seed flour."[164] In view of the threatening atmosphere coloured by the illegally THC drug, one would like to generalise this demand with regard to an independent positive hemp sustainability discussion, which could then also have a retroactive effect on the traditional THC drug perspective; similarly, as this was achieved in 2017 with the medical hemp law, which was immediately followed in May 2017 under the title 'Cannabis and Cannabinoids in Medicine' with a *"further training event of the Cannabis as Medicine Working Group in cooperation with the State Medical Association Hesse and the City of Frankfurt"*[165]

Above all, however, sufficient medical, agricultural and technological research must be developed and promoted for both medical and agricultural hemp; be it within the funding programmes mentioned here occasionally, or be it within the framework of the FONA programme (Research for Sustainability) of the Federal Ministry of Education and Research, in which it *"wants to double research funding for climate protection and more sustainability to four billion euros in the next five years".*[166] With its 2nd goal, 'Explore, protect and use living spaces and natural resources', for example with the sub-goals 'Understand biodiversity changes', 'Secure natural resources' with the *"Actions 14: Stop the pollution of rivers and seas, Action 15: Preserve healthy soils and use land sustainably and Action 16: Further development of agricultural and nutrition systems"*[167] and the chances of a 'circular economy' (Actions 17, 19) there would be enough points of reference.

Such research would not only enable an objective, scientifically supported discussion and enlightenment, but at the same time would also make Germany economically both from an importer to an exporter of high-quality medicine, durable biofibres, building materials such as hemp concrete and insulation materials, in order to thus also drive climate protection worldwide as a pioneer for the so necessary ecological construction and agriculture. (SDG 9, 12–15).

A research that today still leads a rather miserable existence due to the unfortunate fixation on drugs in the discussion. Be it that the Federal Government

---

[164] Pix (2020, p. 9).
[165] Müller-Vahl and Grotenhermen (2017).
[166] BMBF (2020a).
[167] BMBF (2020b).

reacted to the Left's request of 4.7.2019 with the statement: *"Due to the small market volume, any funding programmes are not considered to be sensible."*[168] Or be it that of the 13 research projects listed in the appendix[169] completed between 31.5.2010 and 30.12.2019, only one agricultural project is mentioned, which was quoted above as the 'Winter Hemp' project.[170]

A research situation that is then also reflected in the footnotes of this work as by no means always interest-free sources, which, partly financed with private donations, lack sufficiently tested scientific sources, can occasionally be read more as a utopian hint to possible research approaches.

4.10   SDG 17: Partnerships for Achieving the Goals

The final goal 17 is even more internationally oriented than the rest of Agenda 2030, which is why the Federal Government writes at the beginning of its Sustainability Report 2021: *"The challenges of the world community cannot be tackled alone by governments. The successful implementation of Agenda 2030 therefore requires new forms of cooperation, inter alia, with civil society, national human rights organizations, business and science at local, national and global levels." "In particular, environmentally friendly technologies should be promoted and their dissemination in developing countries should be expanded."*[171]

Leaving aside the question of to what extent the German actors can gain experience from such 'cannabis partnerships', to what extent the trade in medical hemp and industrial hemp can and should be expanded in the future in favor of these developing countries, and whether Germany *"in the field of science, technology and innovation"* will advance cannabis research in the future so that it can offer 'improved access to developing countries', this question of partnership remains a burning issue within our national borders.

As clear as the 'sustainable' benefits of exploiting the cannabis plant are, however, its realization is still difficult, which is practically and ideologically driven or so far mostly prevented by 'political' actors in a wider sense. If one examines this 'political' situation, one repeatedly comes across the problem of the debate

---

[168] Bureg/DIE LINKE (2019, p. 9).

[169] Bureg/DIE LINKE (2019, p. 13–14).

[170] Rinklebe (2019).

[171] DNS Development (2021, p. 335).

on illegalization, in two ways. On the one hand, the discussion of the benefits of hemp always leads back to the field of the THC drug concern dominating the general discussion, such as the regulation of industrial hemp in the BtMG or its connection to the Federal Opium Office. On the other hand, or precisely because of this, the narrower hemp discussion must always push for the legalization of cannabis consumption in order to enable the cultivation of industrial hemp as freely and unhindered as possible. An unholy combination, the solution of which was, for example, a prerequisite for the corresponding economic successes in California[172] or Canada[173] and which was initiated there and here by the 'medical turn'.

In the political realization of these sustainability effects, two fronts are essentially fighting against each other. On the one hand, the representatives of a deeply entrenched drug policy fear, as anchored in the BtMG, used by the political parties in the election campaign and represented by the drug commissioner;[174] and on the other hand, the representatives from science mentioned at the beginning, such as the resolution of criminal law scholars[175] or the reports of the World Commission.[176] A dispute that is also fought out in terms of trade and industrial policy in competition between an emerging hemp industry and an established pharmaceutical or tobacco industry. For the 'hemp side', for example, the agenda of the *Hemp Association* represents the interests of cannabis consumers,[177] and the Cannabis Industry Association (BvCW)[178] or the *European Industrial Hemp Association* (EIHA) represent the producers, while for the medical side[179] the IACM (International Association for Canabionid Medicines)[180] is active.

The actors who occupy this field include the Federal Government in its German Sustainability Strategy 2021: *"To advance the transformations, all actors are needed: • the state and its institutions, • economy, • science and • civil society."*[181] Here, the federal and state governments with their ministries, in particu-

---

[172] Humboldt Seeds (2020).

[173] Government of Canada (2021).

[174] Grauel (2021).

[175] Schildower Kreis (2015).

[176] .Global Commission on Drug Policy (2021).

[177] https://hanfverband.de.

[178] https://start.cannabiswirtschaft.de/

[179] https://eiha.org/

[180] https://www.cannabis-med.org.

[181] DNS Weiterentwicklung (2021, p. 27).

lar justice and the interior, as well as the environment, economy and agriculture, are initially active, for which, since 2017, a "sustainability coordinator at departmental level has been appointed in each ministry as far as possible"[182] A world of state actors that is concretely reflected in legislation, funding programs and, above all, in an ever-expanding bureaucracy.

The actual political parties also play a decisive role, which in their election programs mostly acted on the level of illegalization,[183] and only recently took up the legalization,[184] as well as the audience in the social media, which is almost exclusively concerned with the release of drugs. In the narrower circle of these actors, on the one hand, the scientific media, a multitude of interest groups, such as the BUND, and NGO's as well as the popular media, which, however, can hardly be separated from the pros and cons of this 'drug'.[185]

On the other hand, the market is increasingly taking on a decisive role, on the one hand as a lobby and on the other hand as an example, from the Freiburg pilot project to the international model. In addition to doctors and, initially very restrained, farmers' associations, the direct participants in marketing are mainly fighting each other as lobbyists, for which the BvCW and the European Association EIHA or the Economic Association Cannabis Austria (WCA) are on the one hand and on the other hand those, mostly less openly arguing, established industries. By either earning well on the current medical hemp; so today pharmacies can still demand excessive prices for medical hemp products. Or such lobbyists who fear economic competition of possible hemp products, such as the pharmaceutical industry, but also the established building material and textile industry. But lack of knowledge and economic considerations also prevent the implementation of such alternatives. For example, the representatives of the cement alternative 'Celitement' complain: *"But on a large scale, Celitement, as they call the future cement, is still not produced more than ten years later. But it is scientifically mature,"* says Stemmermann. But the industrial implementation is lengthy. Currently they are working on the realization of their first industrial plant, which is to produce about 50,000 t of cement annually. *"But bringing a new cement to the market is a mammoth task. The search for reliable partners is difficult. And*

---

[182] DNS Weiterentwicklung (2021, p. 90).
[183] Deutscher Hanfverband (2021).
[184] Suliak (2022).
[185] Lesch (2016).

long and expensive approval procedures make it even harder for the research-ers."[186]

Strong partnerships are currently still being organized mainly in the fight for justice by organizations such as the Hemp Association and the research is financed by donations such as the mentioned study by Justus Haucap or the standard application of the youth judge Andreas Müller to the Federal Constitutional Court for the conformity of the cannabis ban. The agenda decision for the lawsuit was financed through donation appeals in Connection of the Justice Offensive of the DHV.[187]

In the field of industrial hemp, the 'European Association for Industrial Hemp (EIHA) u. a. was active with the 'Hemp Manifesto', which was handed over to EU Commission President Ursula von der Leyen on 'Earth Day' 2020, i.e. on 22 April: "*(EIHA) represents the common interests of farmers, producers and traders working with hemp fibers, shives, seeds, leaves and cannabinoids. Our main task is to represent, protect and promote the hemp sector in the EU and in international politics. The EIHA covers various applications of hemp, in particular building materials, textiles, cosmetics, feed, food and dietary supplements.*"[188]

---

[186] Römer (2019).

[187] Deutscher Hanfverband (2020).

[188] EIHA (2020, p. 2).

What is to be Done? 5

5.1 Four Basic Conditions

In order to make this sustainability potential of hemp practically effective, four basic conditions must be implemented more strongly, which would then have to be further concretized for the respective areas.

The first, and probably also decisive condition, is to abandon the close connection to the BtMG, whether through partial or comprehensive legalization of cannabis/hemp, by decisively raising the 0.2% THC limit and/or by transferring the responsibilities from the Ministry of Justice to the Ministry of Agriculture (BMEL). In order to reduce the excessive bureaucracy, the tight cultivation limits and the senseless drug worries.

The second condition, which has been mentioned again and again in the ongoing text, is intensively promoted research, both within the framework of general research funding and through special 'hemp programs': from the breeding of special hemp varieties, such as early-maturing seed hemp, to cultivation, harvesting and processing techniques, especially in the textile, plastic and construction sectors, including the machines required for this, right up to the comparative CO_2 measurement and the medical evaluation research so urgently demanded by the meta-analysis mentioned above. Research that serves both the necessary 'fact-based' enlightenment and the equally necessary cost reduction.

Thirdly, a coordinating program strategy would then stand, which, bundled in one place, develops a hemp sustainability expansion program (HSEP) and implements it politically and financially. After a market analysis and an analysis of the market potential, it would have to develop 'product guidelines' and strategies for tapping the market.[1]

And fourthly, it would be urgently time to also in Germany to tie up an overarching organizational network of those who could be economically interested and involved in the sustainable implementation of these hemp potentials on the 'civil counter side'. For example, following the example of the hemp association carried by consumption-interested people or the industry associations BvCW or Cannabis Austria and EIHA.

5.2   Area-specific Demands

Within this framework, the already formulated area-specific demands could also be incorporated. Thus, the position paper "Cannabis as Medicine" demands, in addition to *"pharmaceutical industry-independent training courses for doctors"* and publicly funded *"clinical research on the effectiveness of cannabis-based medications"*[2] a "reduction of the sales prices for cannabis flowers according to the Schleswig-Holstein model" or the "abolition of the approval requirement of the health insurance companies". And so the Freiburg farmers demand in their paper[3] in addition to "technological and financial investments" *"to fully exploit the ecological and economic potential"* incentives for building with hemp, improved infrastructure and an increase *"in the THC limit of hemp from 0.2% to 5%,* in order to *"be able to cultivate independent of seed merchants, more powerful hemp varieties"*. What the European EIHA concretely further specifies in its hemp manifesto of 2020[4] in ten demands in the economic interest: *"Hemp farmers should receive compensation for the positive external environmental effects, either within the existing or a new emissions trading system"*; *"Promoting the use of hemp-based building materials and other materials [...] in the public and private sectors"* and 'allowing all raw materials obtained from hemp as ingredients for cosmetics'. Finally, 122 criminal law professors demand in their "Resolution"[5] through *"the establishment of an inquiry commission to review the*

---

[1] Bòcsa and Karus, (1997, pp. 370–371).
[2] Stöver et al. (2021, p. 6).
[3] Pix (2020).
[4] EIHA (2020).
[5] Schildower Kreis (2015).

suitability, necessity and normative appropriateness of the narcotics criminal law and, if necessary, to derive proposals for legislative amendments from such an evaluation."

5.3   A 'Sustainable' Conclusion

In summary, from a sustainability perspective, **five priorities emerge for the use of the cannabis plant.**

From a **health perspective**, its medical use is in the foreground, but it is also increasingly used in cosmetic products and food offerings for humans and animals to improve, replace or supplement existing products.

Ecologically, it serves agricultural purposes, such as soil improvement, avoidance of herbicides and pesticides, conservation and improvement of the water balance, and promotion of biodiversity.

Economically, hemp cultivation supports local production, especially in small and medium-sized farms, but also opens up new forms of trade and industry at the national and international level.

In the context of the increasing climate crisis, it helps above all to cope with the CO_2**problem.** On the one hand, it binds CO_2 similarly to afforestation, and on the other hand, it serves as a raw material replacement, for example in clothing, technology (automobile construction) or in the construction industry.

Under the general aspect *"creating more inclusive and fairer societies everywhere"*, the—not further treated here—**legalization** of cannabis should not only correspond to the demands of the human rights convention, but also save considerable financial and humanitarian costs.

The chances of a sustainable use of hemp should be driven forward in politics and by the German citizen. At the moment, the immense prices and the high demand unfortunately show a different picture of the citizen who wants but cannot or may not, of farmers who have to overcome immense legal hurdles, and may not use all parts of the plant, of an industry that is deterred from investing by complex rules and laws, up to a society that is increasingly threatened by a growing climate crisis.

The use of industrial hemp in the context of the UN Agenda 2030 could make this hemp a big player in the field of sustainability, provided it is allowed to unfold its potentials, as its components from seed to fiber and shives to root, its cultivation and its utilization can directly and indirectly make a not insignificant contribution to their SDG sustainability goals.

If Germany were to embark on this sustainability path, as is now often demanded, it would not only become economically independent, for example in the field of oil-based raw materials, but also, through the political and technological innovations required for this, 'exemplarily' promote the concept of Agenda 2030, which has been demanding a more sustainable world since 2015.

What You Can Take Away From This *essential*

- An insight into the current discourse on sustainability
- An understanding of the barriers to the ongoing legalization/illegalization of hemp and the role of the stakeholders involved in the debate
- The meaning of relevant terms such as phytosanitation, cottonization, urban mining, cascade utilization and circular economy
- References from government publications, media and market offerings

References

Abbott, Graham (2020): Hemp-Based 'Cannabis Car' is Carbon Neutral, Stronger than Steel, in: Ganjapreneur, 18.05.2020, https://www.ganjapreneur.com/carbon-neutral-hemp-based-cannabis-car/ (accessed: 26.12.2021)

AEPW (2019): Alliance to End Plastic Waste. BASF, Henkel & Co. found alliance against plastic waste: in: markenartikel-magazin.de, 17.01.2019, https://www.markenartikel-magazin.de/_rubric/detail.php?rubric=marke-marketing&nr=24583&PHPSESSID=lg78fem7bi2nfutfvuthh1fcu0 (accessed: 26.12.2021)

Akzept/Aidshilfe (2019): Bundesverband Akzept e. V.; Deutsche Aids-Hilfe. 6. Alternativer Drogen- und Suchtbericht. Pabst Science Publishers (Lengerich)

Akzept/Aidshilfe (2020): Bundesverband Akzept e. V.; Deutsche Aids-Hilfe. 7. Alternativer Drogen- und Suchtbericht. Pabst Science Publishers (Lengerich)

Akzept/Aidshilfe (2021): Bundesverband Akzept e. V.; Deutsche Aids-Hilfe. 8. Alternativer Drogen- und Suchtbericht. Pabst Science Publishers (Lengerich)

Aposcope (2022): Zukunftsmarkt Cannabis. Insights aus der Apotheke 2022 https://marktforschung.aposcope.de/zukunftsmarkt-medizinisches-cannabis-2022/ (accessed: 05.07.2022)

Barrett, John (2020): Ecological footprint and water analysis of cotton, hemp and polyester, in: SEI, 11.03.2020, https://www.sei.org/publications/ecological-footprint-water-analysis-cotton-hemp-polyester/ (accessed: 23.11.2021)

BBSR (2016): Federal Institute for Research on Building, Urban Affairs and Spatial Development: Data basis for the building stock https://www.bbsr.bund.de/BBSR/DE/veroeffentlichungen/analysen-kompakt/2016/ak-09-2016-dl.pdf?__blob=publicationFile&v=2 (accessed: 26.12.2021)

BfArM (2020): Cannabis Agency, in: Federal Institute for Drugs and Medical Devices https://www.bfarm.de/DE/Bundesopiumstelle/Cannabis-als-Medizin/Cannabisagentur/_node.html (accessed: 22.12.2021)

BfArM (2021): BfArM starts sales of cannabis for medical purposes to pharmacies: in: Federal Institute for Drugs and Medical Devices, 07.07.2021, https://www.bfarm.de/SharedDocs/Pressemitteilungen/DE/2021/pm6-2021.html (accessed: 22.12.2021)

BfArM (2022a): Cannabis Agency. https://www.bfarm.de/DE/Bundesopiumstelle/Cannabis-als-Medizin/Cannabisagentur/_node.html (accessed: 05.07.2022)

© The Editor(s) (if applicable) and The Author(s), under exclusive license to Springer Fachmedien Wiesbaden GmbH, part of Springer Nature 2023
J. M.W. Westphal, *The Sustainability of Hemp,* Springer essentials, https://doi.org/10.1007/978-3-658-41819-9

BfArM (2022a): Cannabis as medicine: BfArM publishes final report on accompanying survey. Number 5/22 of 06.02.2022b. https://www.bfarm.de/SharedDocs/Pressemitteilungen/DE/2022/pm05-2022.html (accessed: 05.07.2022)

BFN (no date): Federal Agency for Nature Conservation: The Convention on Biological Diversity (CBD) https://www.bfn.de/das-uebereinkommen-ueber-die-biologische-vielfalt-cbd (accessed: 27.12.2021)

BLE-Sortenliste (2022): Federal Agency for Agriculture and Food: (BLE) Variety List https://www.ble.de/SharedDocs/Downloads/DE/Landwirtschaft/Nutzhanf/Sortenliste.pdf?__blob=publicationFile&v=8 (accessed: 05.07.2022)

BLE 2021a: Federal Agency for Agriculture and Food: (BLE) (1.10.2021). Press release— Hemp cultivation 2021: Number of farms and area continued to grow (https://www.ble.de/SharedDocs/Pressemitteilungen/DE/2021/211001_Nutzhanfanbau.html (accessed: 05.07.2022)

BLE (2022): Federal Agency for Agriculture and Food: (BLE). INFORMATION SHEET For farmers who will grow hemp in 2022. https://www.ble.de/SharedDocs/Downloads/DE/Landwirtschaft/Nutzhanf/MerkblattLandwirte.pdf;jsessionid=8600386AF2C447F1D3C88753483735B3.1_cid325?__blob=publicationFile&v=9 (zugegriffen: 05.07.2022)

BLE Nutzhanf (o. J.): Bundesanstalt für Landwirtschaft und Ernährung: Nutzhanf, in: BLE, o. D., https://www.ble.de/DE/Themen/Landwirtschaft/Nutzhanf/nutzhanf_node.html (zugegriffen: 26.12.2021)

Blienert, Burkart: (2021): Cannabis im Koalitionsvertrag: Wir schreiben europäische Geschichte: In: https://www.vorwaerts.de/artikel/cannabis-koalitionsvertrag-schreiben-europaeische-geschichte (zugegriffen: 05.07.2022)

BMBF (2020a): Bundesministerium für Bildung und Forschung: Forschung für Nachhaltigkeit, in https://www.bmbf.de/bmbf/de/forschung/umwelt-und-klima/forschung-fuer-nachhaltigkeit/forschung-fuer-nachhaltigkeit.html (zugegriffen: 26.12.2021)

BMBF (2020b): Federal Ministry of Education and Research. SDG Goals, Research for Sustainability, in: Fona https://www.fona.de/medien/pdf/Ziele-FONA-Strategie.pdf (accessed: 26.12.2021)

BMEL (2021): Federal Ministry of Food and Agriculture: Areas for the raw materials of the future, in: BMEL, o. D., https://www.bmel.de/DE/themen/landwirtschaft/bioeokonomie-nachwachsende-rohstoffe/nachwachsende-rohstoffe-flaechen.html (accessed: 23.03.2021)

BMEL (2022a): Federal Ministry of Food and Agriculture: Agriculture, climate protection and climate resilience in: BMEL, 22.06 2022, https://www.bmel.de/DE/themen/landwirtschaft/klimaschutz/landwirtschaft-und-klimaschutz.html (accessed: 05.07.2022)

BMEL (2022a): Federal Ministry of Food and Agriculture: Joint task "Improvement of agricultural structures and coastal protection", in: BMEL 21. Feb. 2022b, https://www.bmel.de/DE/themen/laendliche-regionen/foerderung-des-laendlichen-raumes/gemeinschaftsaufgabeagrarstrukturkuestenschutz/gak.html;jsessionid=B019872672DF0C89CA015A83D47039EB (accessed: 05.07.2022)

BMI (2021): Federal Ministry of the Interior and Home Affairs: BMI promotes climate protection research in the building sector with the Future Building Innovation Programme, in: Federal Ministry of the Interior and Home Affairs, 16.02.2021, https://

www.bmi.bund.de/SharedDocs/pressemitteilungen/DE/2021/02/klimaschutzforschung-im-gebaeudebereich.html (accessed: 26.12.2021)

BMUV (2021): Federal Ministry for the Environment, Nature Conservation, Nuclear Safety and Consumer Protection. Press release no. 230/21 of 8.9.2021: Better protection for insects through less use of plant protection products. https://www.bmuv.de/pressemitteilung/besserer-schutz-fuer-insekten-durch-weniger-pflanzenschutzmittel-einsatz (accessed: 27.12.2021)

BMWI (2014): Federal Ministry of Economics and Energy: Sanitation needs in the existing building stock, https://www.bmwi.de/Redaktion/DE/Publikationen/Energie/sanierungsbedarf-im-gebaeudebestand.pdf?__blob=publicationFile&v=3 (accessed: 25.12.2021)

BMWI (2021): Federal Ministry of Economics and Energy: Transport in: BMWI, 19.01.2021, https://www.bmwi.de/Redaktion/DE/Dossier/leichtbau.html (accessed: 26.12.2021)

Bócsa, Iván; Karus, Michael (1997): Hanfanbau. Botanik, Sorten, Anbau und Ernte. C.F. Müller

Boedefeld, Christian (2021): Deutscher Großhändler importiert Cannabis aus Lesotho, in: Hanf Magazin, 10.11.2021, https://www.hanf-magazin.com/news/deutscher-grosshaendler-importiert-cannabis-aus-lesotho/ (accessed: 22.12.2021)

Böllinger, Lorenz (2016): Freigabe (in) der Diskussion. In: Helmut Pollähne, Christa Lange-Joest (Hg.): Rauschzustände. Drogenpolitik – Strafjustiz – Psychiatrie. LIT-Verlag (Berlin) 2016: 89–112

Böhning-Gaese, Katrin (2021): Der Verlust der Biodiversität und was wir tun können, in: Forschung und Lehre, 06.05.2021, https://www.forschung-und-lehre.de/zeitfragen/derverlust-der-biodiversitaet-und-was-wir-tun-koennen-3698/ (accessed: 26.12.2021)

Brandt, Mathias (2018): Medizinisches Cannabis in Deutschland, in: Statista Infografiken, 30.11.2018, https://de.statista.com/infografik/9533/verwendung-von-medizinischemcannabis-in-deutschland/ (accessed: 25.12.2021)

Braun, Reiner (2020): Housing Market Forecast 2021/22, in: www.empirica-institut, 08.2020, https://www.empiricainstitut.de/fileadmin/Redaktion/Publikationen_Referenzen/PDFs/empi256rb.pdf (accessed: 26.12.2021)

Brosius, Emily Gray (2016): Italian Farmers are planting Hemp to clean Polluted Soil. https://netzfrauen.org/2016/07/26/italien-hanf/ (accessed: 05.07.2022)

Brundtland Report (1987): https://www.are.admin.ch/are/de/home/medien-und-publikationen/publikationen/nachhaltige-entwicklung/brundtland-report.html (accessed: 05.07.2022)

Federal Constitutional Court (2021): BUND for Nature Conservation and the Environment in Germany, 29.04.2021, https://www.bund.net/service/presse/pressemitteilungen/detail/news/bahnbrechendes-klima-urteil-des-bundesverfassungsgerichts/ (accessed: 20.10.2021)

Bureg (2019a): Bundesregierung (2019): Insekten besser schützen. www.bundesregierung.de/breg-de/aktuelles/aktionsprogramm-insektenschutz-1581358 (zugegriffen: 05.07.2022)

Bureg (2020a): Bundesregierung: Mehr E-Mobilität, in: Bundesregierung, 2020, https://www.bundesregierung.de/breg-de/themen/klimaschutz/verkehr-1672896 (zugegriffen: 28.12.2021)

Bureg (2021): Bundesregierung, 04.07.2021: Einweg-Plastik wird verboten.

https://www.bundesregierung.de/breg-de/themen/nachhaltigkeitspolitik/einwegplastik-wird-verboten-1763390 (zugegriffen: 26.12.2021)

Bureg (2022): Bundesregierung, 24.2.2022: Insektenschutz. Weniger Pflanzenschutzmittel einsetzen. https://www.bundesregierung.de/breg-de/suche/insekten-schuetzen-1852558 (zugegriffen: 05.07.2022)

Bureg (2022a): Bundesregierung, 17.2.2022. Mehr bezahlbare und klimagerechte Wohnungen schaffen https://www.bundesregierung.de/breg-de/suche/wohnungsbau-bundesregierung-2006224 (zugegriffen: 05.07.2022)

Bureg/BÜNDNIS 90/DIE GRÜNEN (2020): (Antwort der Bundesregierung auf die Kleine Anfrage der Abgeordneten Dr. Kirsten Kappert-Gonther, Maria Klein-Schmeink, Kordula Schulz-Asche, weiterer Abgeordneter und der Fraktion BÜNDNIS 90/DIE GRÜNEN) Drucksache 19/22651, in: Deutscher Bundestag, 17.09.2020, https://dserver.bundestag.de/btd/19/226/1922651.pdf (zugegriffen: 15.10.2021)

Bureg/DIE LINKE (2019b): Antwort der Bundesregierung auf die Anfrage der Abgeordneten Kirsten Tackmann, Niema Movassat, Gesine Lötzsch: Nutzhanf- Agrarstoff mit Potential, in: Deutscher Bundestag, 04.07.2019, https://dserver.bundestag.de/btd/19/113/1911377.pdf (zugegriffen: 22.12.2021)

Bureg/DIE LINKE (2020b): Antwort der Bundesregierun: Versorgungssituation und Bedarf von medizinischem Cannabis, in: Deutscher Bundestag, 23.03.2020, https://dserver.bundestag.de/btd/19/182/1918292.pdf (zugegriffen: 16.11.2021)

Bureg/DIE LINKE/Bündnis 90/DIE GRÜNEN (2021): Response of the Federal Government: Make full use of the potential of industrial hemp cultivation, in: German Bundestag, 14.01.2021, https://dserver.bundestag.de/btd/19/258/1925883.pdf (accessed: 22.12.2021)

Bureg/FDP (2020c): Response of the Federal Government to the request of the Members of Parliament Nicole Bauer, Frank Sitta, Dr. Gero Clemens Hocker, other Members of Parliament and the FDP Parliamentary Group. Drucksache 19/24964 -Potential of the hemp plant-, in: German Bundestag, 21.12.2020, https://dserver.bundestag.de/btd/19/254/1925497.pdf (accessed: 25.12.2021)

Business Insider Deutschland (2021): Legalization of cannabis: Özdemir announces large-scale cultivation, in: Business Insider, 27.12.2021, https://www.businessinsider.de/politik/legalisierung-von-cannabis-bauern-in-den-startloechern-a/ (accessed: 28.12.2021)

BvCW Elemente Bd. 3 (2021a): Cannabis Industry Association e. V. Elemente Band 3: Positions & Goals
of the Medical Cannabis Department. In: https://start.cannabiswirtschaft.de/wp-content/uploads/2021/10/ELEMENTE_3_Positionen_Medizinalcannabis_BvCW.pdf (accessed: 05.07.2022)

BvCW. Elemente Bd. 4 (2020): Brachenverband Cannabiswirtschaft e. V. Elemente Band 4: For a regulated CBD market! https://start.cannabiswirtschaft.de/wp-content/uploads/2021/02/ELEMENTE_4_Positionierung_CBD_BvCW.pdf (accessed: 05.07.2022)

BvCW. Elemente Bd 5: Branchenverband Cannabiswirtschaft e. V. Elemente Band 5: Overview of the legal status of CBD in Europe (accessed: 05.07.2022)

BvCW. Elemente Bd. 12 (2021): Branchenverband Cannabiswirtschaft e. V. Elemente Band 12: Hemp as a renewable raw material. Positions and demands—from the field of industrial hemp & food. In: https://start.cannabiswirtschaft.de/wp-content/

uploads/2021/05/ELEMENTE12_Nutzhanf_Positionierungen_BvCW.pdf   (accessed: 05.07.2022)

BvCW Elemente Bd. 15 (2021b): Branchenverband Cannabiswirtschaft e. V. Elemente Band 15: On the handling of cannabis and cannabis products in the EU. Translation of an overview of legal requirements https://start.cannabiswirtschaft.de/wp-content/uploads/2021/08/ELEMENTE_15_Cannabis_in_der_EU_BvCW.pdf   (accessed: 05.07.2022)

BvCW Elemente Bd. 19 (2022): Branchenverband Cannabiswirtschaft e. V. Elemente Band 19: Industrial hemp in Germany—overview in numbers: In: https://start.cannabiswirtschaft.de/wp-content/uploads/2022/03/ELEMENTE_19_V1.2_Zahlenwerk_Nutzhanf_BvCW.pdf (zugegriffen: 05.07.2022)

Callaway, J. (2004): Hempseed as a nutritional resource: An overview, in: SpringerLink, 01.01.2004, https://link.springer.com/article/https://doi.org/10.1007/s10681-004-4811-6?error=cookies_not_supported&code=4d365ef0-f7f5-46e7-ae9d-b02d31bd15a5 (zugegriffen: 25.12.2021)

Canna Connection (2021): Chernobyl – Sorteninformationen: in: Cannaconnection, https://www.cannaconnection.de/sorten/chernobyl (zugegriffen: 26.12.2021)

Cantourage (2021): medizinisches Cannabis aus Jamaika in Deutschland ab sofort verfügbar, in: Krautinvest, 01.09.2021, https://krautinvest.de/cantourage-cannabis-aus-jamaika-in-deutschland-ab-sofort-verfuegbar/ (zugegriffen: 22.12.2021)

CANSATIVA (2021): Medizinisches Cannabis für Apotheken und Großhandel: in: Cansativa, 07.07.2021, https://www.cansativa.de/de/ (zugegriffen: 22.12.2021)

Civantos D. (2017): Cannabis can detoxify and regenerate contaminated soils https://www.dinafem.org/de/blog/cannabis-aufbereitung-verseuchte-boden/ (accessed: 05.07.2022)

Demortain, Pierre (2018): Breakthrough in lightweight biomaterials gains momentum: in: Faurecia, 14.03.2018, https://www.faurecia.com/en/newsroom/breakthrough-lightweight-biomaterials-gains-momentum?fbclid=IwAR0OMAHQAky4_kRPwcGYZ-A5yh18GJP3NhupYqW_juvxbKt28fstAQ2tnovs (accessed: 26.12.2021)

Deutscher Hanfverband (2020): Federal Constitutional Court to review cannabis ban, in: Deutscher Hanfverband, 03.06.2020, https://hanfverband.de/nachrichten/pressemitteilungen/bundesverfassungsgericht-vor-pruefung-des-cannabisverbots   (accessed: 27.12.2021)

Deutscher Hanfverband (2021): Election analysis for the federal election on 26.09.2021 https://hanfverband.de/wahlcheck_btw21 (accessed: 27.12.2021)

Deutsches Institut für Entwicklungspolitik: Sustainable Development Solutions Network (SDSN), in: The German Development Institute/Deutsches Institut für Entwicklungspolitik (DIE), o. D., https://www.die-gdi.de/sdsngermany/ (accessed: 19.10.2021)

Dinafem Seeds (2021): Buy cannabis seeds of the highest quality, in: Dinafem, https://www.dinafem.org/de/hanfsamen-kaufen/ (accessed: 26.12.2021)

DNS (2017): German Sustainability Strategy Revised Edition 2016. In: Sustainability Report, 2017, https://www.bundesregierung.de/resource/blob/975292/730844/3d30c6c2875a9a08d364620ab7916af6/deutsche-nachhaltigkeitsstrategie-neuauflage-2016-download-bpa-data.pdf (accessed: 05.07.2022)

DNS Draft (2021): German Sustainability Strategy – Draft https://www.bundesregierung.de/resource/blob/998006/1793018/73d3189a28be9f3043c7736d3e1de4df/dns2021-dialogfassung-data.pdf?download=1 (accessed: 05.07.2022)

DNS Development (2021): German Sustainability Strategy Development in: www.bundesr-egierung.de, 03.08.2021, https://www.bundesregierung.de/resource/blob/998006/187-3516/3d3b15cd92d0261e7a0bcdc8f43b7839/2021-03-10-dns-2021-finale-langfassung-nicht-barrierefrei-data.pdf?download=1 (accessed: 07.10.2021)

Drug Commissioner (2021): Press release of the Drug Commissioner of 27.7.2021 Drug-related crime in Germany continues to rise in: The Drug Commissioner of the Federal Government, 27.07.2021, https://www.drogenbeauftragte.de/presse/detail/rauschgift-kriminalitaet-in-deutschland-steigt-weiter-an/ (accessed: 15.10.2021)

Dutta, Dipayan (2018): Forget Electric Cars! Henry Ford's Cannabis car was made from Hemp: 10xStronger than steel, 100% green!, in: The Financial Express, 17.11.2018, https://www.financialexpress.com/auto/car-news/forget-electric-cars-henry-fords-can-nabis-car-was-made-from-hemp-10xstronger-than-steel-100-green/1384733/ (accessed: 26.12.2021)

EIHA (2020a): Hanf ein wirklich grüner Deal, in: European Industrial Hemp Association https://eiha.org/wp-content/uploads/2020/09/Hanf-ein-wirklicher-gru%CC%88ner-Deal_DE.pdf (accessed: 25.12.2021)

EIHA (2020a): EIHA publishes hemp manifesto: hemp as a guide to sustainable economy, in: presseportal.de, 21.04.2020b, https://www.presseportal.de/pm/141925/4576796 (accessed: 27.12.2021)

Endris, Julia (2020): Medizinalhanf: Wer beliefert die Apotheken mit deutschem Can-nabis?, in: Avoxa – Mediengruppe Deutscher Apotheker GmbH, 20.08.2020, https://www.pharmazeutische-zeitung.de/wer-beliefert-die-apotheken-mit-deutschem-canna-bis-119659/ (accessed: 22.12.2021)

Energie-Experten (2016): Dämmung mit Hanf: Herstellung, Dämmwerte und Verarbei-tung im Überblick, in: energie-experten, 08.06.2016, https://www.energie-experten.org/bauen-und-sanieren/daemmung/daemmstoffe/hanfdaemmung (accessed: 25.10.2021)

Faurecia (2018): Immer mehr Biomaterialien im Auto: Faurecia setzt seit 10 Jahren auf NafiLean: in: Wiztopic, 21.07.2018, https://www.wiztopic.com/news/immer-mehr-bio-materialien-im-auto-faurecia-setzt-seit-10-jahren-auf-nafileantm-und-treibt-nachhaltige-mobilitatslosungen-voran-77e7-0a54a.html (accessed: 26.12.2021)

FAO (2020): Food and Agriculture Organization of the United Nations: The Potential of Agroecology to Secure Against Climate Change and Build Resilient and Sustain-able Livelihoods and Food Systems, in: FAO, 2020, http://www.fao.org/3/cb0486de/CB0486DE.pdf (accessed: 26.12.2021)

Flicker, Nathaniel; K. Poveda; H. Grab (2020): The Bee Community of Cannabis sativa and Corresponding Effects of Landscape Composition, in: Environmental Entomology Oxford Academic, 02.2020, https://academic.oup.com/ee/article/49/1/197/5634339?logi n=true (accessed: 27.12.2021)

FNR (o. J.): Fachagentur Nachwachsende Rohstoffe: Winterhanf (Hanf als Winterzwis-chenfrucht). https://pflanzen.fnr.de/industriepflanzen/faserpflanzen/winterhanf-hanf-als-winterzwischenfrucht (accessed: 27.12.2021)

FNR (2022). Fachagentur Nachwachsende Rohstoffe: Förderprogramm „Nachwachsende Rohstoffe", in: FNR, o. D., https://www.fnr.de/projektfoerderung/foerderprogramm-nachwachsende-rohstoffe (accessed: 05.07.2022)

Frahm, Christian (2013): Hanffasern im Autobau, in: DER SPIEGEL, 14.03.2013, https://www.spiegel.de/auto/aktuell/autos-aus-hanf-naturfasern-werden-in-der-karosserie-verbaut-a-878973.html (accessed: 26.12.2021)

Gassmann, Michael (2020): So will die Anti-Müll-Allianz die Welt vom Plastik befreien. In: https://www.welt.de/wirtschaft/article214749950/Allianz-gegen-Plastikmuell-in-der-Umwelt-1-5-Milliarden-Dollar-im-Kampf-gegen-Abfall.html (accessed: 05.07.2022)

Gebhardt, Kathrin (2016): Backen mit Hanf: Berauschend gut!, 3. Aufl., Aarau, Schweiz: Nachtschatten Verlag

Geyer, Steffen (2006): Warum Hanf? Über die ökologischen und ökonomischen Möglichkeiten des Biorohlstoffs Hand. In; Hanfverband 2006. https://hanfverband.de/sites/hanfverband.de/files/dhv_warum_hanf.pdf (accessed: 16.11.2021)

Global Commission on Drug Policy (2021): in: The Global Commission on Drug Policy, Reports. https://www.globalcommissionondrugs.org/reports (accessed: 27.12.2021)

Götze, Susanne: (2021): Many existing buildings are a disaster, in: DER SPIEGEL, Hamburg, Germany, 31.08.2021, https://www.spiegel.de/wissenschaft/mensch/klimaschutz-bei-gebaeuden-viele-bestandsbauten-sind-fuers-klima-eine-katastrophe-a-60b0e8c4-9256-472a-888c-ec6f531683d6 (accessed: 16.11.2021)

Government of Canada (2021): Cannabis Legalization and Regulation: in: Government of Canada, 07.07.2021, https://www.justice.gc.ca/eng/cj-jp/cannabis/ (accessed: 27.12.2021)

Grauel, Marie-Luise (2021): Drugs Commissioner: Possession of six grams of cannabis should no longer be a criminal offence, in: Berliner Zeitung, 23.08.2021, https://www.berliner-zeitung.de/news/bundesdrogenbeauftragte-6-gramm-cannabis-besitz-soll-keine-straftat-mehr-sein-li.178462 (accessed: 25.10.2021)

Greenpeace (2019): Climate killer plastic, in: Greenpeace, 17.05.2019, https://www.greenpeace.de/engagieren/nachhaltiger-leben/klimakiller-kunststoff (accessed: 26.12.2021)

Greenpeace (2020): The bill comes due, in: Greenpeace, 10.04.2020, https://www.greenpeace.de/biodiversitaet/meere/meeresschutz/rechnung-geht (accessed: 26.12.2021)

Greenvision (2019): Hemp plastic as a packaging alternative, in: CBD Guide Austria, 18.07.2019, https://cbdguideaustria.com/hanfplastik-als-verpackungsalternative/ (accessed: 26.12.2021)

Grimm, Roland (2020): Eco-wall material: What is hemp lime?, in: baustoffwissen, 03.02.2020, https://www.baustoffwissen.de/baustoffe/baustoffknowhow/boden_und_wand/oeko-wandbaustoff-was-ist-hanfkalk/ (accessed: 25.12.2021)

Grotenhermen, Franjo (2020): The healing power of CBD and cannabis: How hemp products can improve our health. Rowohlt Verlag

Grotenhermen, Franjo (2021): Cannabis as medicine: Irritating statements in the Cannabis Report of the BKK Mobil Oil. In: akzept e. V., Deutsche Aidshilfe (eds) 8. Alternativer Drogen- und Suchtbericht 2021 (Akzept/Aidshilfe (2021 S. 155–162)

Handelsblatt (2018): In Canada, the medical cannabis market is booming. In: Handelsblatt from 19.2.2018 https://www.handelsblatt.com/unternehmen/handel-konsumgueter/pharmaindustrie-in-kanada-boomt-der-markt-fuer-medizinisches-cannabis/20979060.html (accessed: 05.07.2022)

https://www.handelsblatt.com/unternehmen/handel-konsumgueter/pharmaindustrie-in-kanada-boomt-der-markt-fuer-medizinisches-cannabis/20979060.html?ticket=ST-6884829-esWXgDxdin5ml19gbrfK-cas01.example.org (accessed: 25.12.2021)

Hanfare (o. J.) Your reliable mail-order house for hemp products since 2005. In, https://www.hanfare.de/?gclid=Cj0KCQjwm9yJBhDTARIsABKIcGaIvbSRE414jkhv9H BugBc39GK9jnAa9WHSX0tLgxPHZKnJ4D0fN7AaAuJ9EALw_wcB (accessed: 25.12.2021)

Haucap, Justus (2018): The costs of cannabis prohibition, in: Hanfverband, 14.11.2018, https://hanfverband.de/sites/default/files/cannabis_final-141118.pdf (accessed: 27.12.2021)

Haucap, Justus; Knoke, Leon (2021): Fiscal effects of cannabis legalization in Germany: An update. https://www.dice.hhu.de/fileadmin/redaktion/Fakultaeten/Wirtschaftswissenschaftliche_Fakultaet/DICE/Bilder/Nachrichten_und_Meldungen/Fiskalische_ Effekte_Cannabislegalisierung_final.pdf (accessed: 05.07.2022)

Hempopedia (o. J.a): Use in the automotive industry. In: http://www.hempopedia.com/verwendungundverbreitung/verwendunginderautomobilindus.html (accessed: 26.12.2021)

Hempopedia (o. J.b): Use in the paper industry. In: http://www.hempopedia.com/verwendungundverbreitung/hanfinderpapierindustrie.html (accessed: 26.12.2021)

HempToday (2020): European Parliament signs off on raising EU THC limit to 0.3%: in: HempToday, 30.10.2020, https://hemptoday.net/european-parliament-signs-off-on-raising-eu-thc-limit-to-0-3/ (accessed: (22.12.2021)

Herer, Jack und M. Bröckers (1994): Die Wiederentdeckung der Nutzpflanze Hanf. Heyne-Verlag

Hoch, Eva; Chris Maria Friemel; Miriam Schneider (2019): Cannabis: Potenzial und Risiko: Eine wissenschaftliche Bestandsaufnahme, 1. Aufl. 2019, Berlin, Deutschland: Springer

Humboldt Seeds (2017): Von Apulien nach Tschernobyl: Die Wirkung von Hanf auf die Regeneration des Erdreichs, in: Humboldt Seeds, 06.01.2017, https://www.humboldtseeds.net/de/blog/cannabis-zur-regenerierung-von-boeden/ (zugegriffen: 26.12.2021)

Humboldt Seeds (2020a): Zehn große Probleme der Marihuanaindustrie zwei Jahre nach der Legalisierung: in: Humboldt Seeds, 14.02.2020, https://www.humboldtseeds.net/ de/blog/probleme-kalifornische-marihuanaindustrie-nach-legalisierung/ (zugegriffen: 27.12.2021)

IPCC (2021): AR6 Climate Change 2021: The Physical Science Basis: in: IPCC, 06.08.2021, https://www.ipcc.ch/report/sixth-assessment-report-working-group-i/ (zugegriffen: 18.10.2021)

IPCC (2022): IPCC sixth Assessment Report. Mitigation o Climate Change. https://www.ipcc.ch/report/ar6/wg3/ (zugegriffen: 05.07.2022)

Jacobs, Caleb (2017): The Renew SportsCar is Made With 100 Pounds of Cannabis, in: The Drive, 20.07.2017, https://www.thedrive.com/article/12712/the-renew-sports-car-is-made-with-100-pounds-of-cannabis (zugegriffen: 26.12.2021)

Jähnert, Christopher (2021): Christopher Jähnert im Interview mit der Drogenbeauftragten Daniela Ludwig: „Sehe keinen Weg zur Cannabis-Legalisierung", in: swr.online, 11.06.2021, https://www.swr.de/swraktuell/radio/drogenbeauftragte-sehe-keinen-weg-zur-cannabis-legalisierung-100.html (zugegriffen: 17.10.2021)

Johnsen, Erik und G. Maag (2020): Medizinal-Cannabis, in: IQVIA – Powering Healthcare with Connected Intelligence, 2020, https://www.iqvia.com/-/media/iqvia/pdfs/germany/ publications/artikel-in-der-fachpresse/iqvia-artikel-medizinal-cannabis-pharmind-0621. pdf (zugegriffen: 25.12.2021)

Karlsson, L.; Finell, M.; Martinson, K. (2010): Effects of increasing amounts of hemp-seed cake in the diet of dairy cows on the production and composition of milk: in: ScienceDirect, 01.01.2010, https://www.sciencedirect.com/science/article/pii/S1751731110001254 (zugegriffen: 25.12.2021)

Klein, Axel und B. Stothard (2018): Collapse of the Global Order on Drugs: From UNGASS 2016 to Review 2019. Emerald Publishing Limited

Klimapakt Deutschland (2021): in: Bundesministerium für Umwelt, Naturschutz und nukleare Sicherheit, 12.05.2021, https://www.bmu.de/fileadmin/Daten_BMU/Download_PDF/Klimaschutz/klimapakt_deutschland_bf.pdf (zugegriffen: 20.10.2021)

Klimaschutzgesetz 2021: in: Generationenvertrag für das Klima, 12.07.2021, https://www.bundesregierung.de/breg-de/themen/klimaschutz/klimaschutzgesetz-2021-1913672 (zugegriffen: 21.10.2021)

Klima-Übereinkommen von Paris (2015): ÜBEREINKOMMEN VON PARIS: in: EUR-Lex Access to European Union Law, 19.10.2016, https://eur-lex.europa.eu/legal-content/DE/TXT/?uri=CELEX:22016A1019(01) (zugegriffen: 16.10.2021)

Klöckner, Julia (2018): Neue Produkte: Aus Natur gemacht, in: Bundesministeriums für Ernährung und Landwirtschaft, 03.08.2018, https://www.bmel.de/SharedDocs/Downloads/DE/Broschueren/NeueProdukteAusNaturGemacht.pdf?__blob=publicationFile&v=8 (zugegriffen: 22.12.2021)

Knauer, Sebastian (2005): Banane im Heck. https://www.spiegel.de/auto/werkstatt/daimler-chrysler-banane-im-heck-a-363551.html (zugegriffen: 05.07.2022)

Knodt, Micha (2021): Medizin, Droge oder Lebensmittel? Die EU streitet seit Jahren um CBD, in: Krautinvest, 05.05.2021, https://krautinvest.de/medizin-droge-oder-lebensmittel-die-eu-streitet-seit-jahren-um-cbd/ (zugegriffen: 25.10.2021)

kochbar (2022): https://www.kochbar.de/rezepte/hanf.html (zugegriffen: 05.07.2022)

Kuebler, Martin (2020): Wie der Klimawandel die Landwirtschaft in Europa verändert. https://www.dw.com/de/klimawandel-dürre-landwirtschaft-deutschland-bauer-trockenheit/a-53825523 (zugegriffen: 05.07.2022)

Lesch, Harald (2016): Wie gefährlich ist Hanf?, in: YouTube, 11.05.2016, https://www.youtube.com/watch?v=FD1HikaELK4 (accessed: 27.12.2021)

MRL Baden-Württemberg (2020): Ministry for Rural Areas and Consumer Protection Baden-Württemberg: Information sheet for the application for direct payments for hemp areas, in: Agrarpolitik & Förderung, 03.2020, https://foerderung.landwirtschaft-bw.de/pb/site/pbs-bw-mlr/get/documents_E-126616404/MLR.LEL/PB5Documents/fiona/2020/Merkblaetter/DZ%20-%20Merkblatt%20zum%20Anbau%20von%20Hanf%20GA%202020.pdf?attachment=true (accessed: 27.12.2021)

Müller-Vahl, Kirsten; Grotenhermen, Franjo (2017): Medical Cannabis: The Most Important Changes. Dtsch Arztebl 2017; 114(8): A-352/B-306/C-300. http://politik-fuer-menschen-mit-handicap.de/documents/Deutsches_Aerzteblatt_Medizinisches_Cannabis_-_Die_wichtigsten%20Aenderungen_(24.02.2017).pdf

Neijat M. et al (2015): Hempseed Products Fed to Hens Effectively Increased n-3 Polyunsaturated Fatty Acids in Total Lipids, Triacylglycerol and Phospholipid of Egg Yolk, in: SpringerLink, 29.10.2015, https://link.springer.com/article/https://doi.org/10.1007/s11745-015-4088-7?error=cookies_not_supported&code=a7d9f330-7023-4217-b133-8bc9e2bb7c83 (accessed: 25.12.2021)

Nette-group (2015): Hanf in der Textilindustrie, in: NG, 2015, https://nette-group.de/was-ist-hanf/hanf-als-rohstoff/hanf-in-der-textilindustrie (zugegriffen: 25.12.2021)

Omar, Faruk et al (2012): Biocomposites reinforced with natural fibers, in: ScienceDirect, 01.11.2012, https://www.sciencedirect.com/science/article/pii/S0079670012000391 (zugegriffen: 26.12.2021)

Orlowicz, Jessica (2020): Bienen fliegen auf Cannabis: Hanf könnte dem Bienensterben entgegenwirken, in: RND, 26.07.2020, https://www.rnd.de/wissen/bienen-fliegen-auf-cannabis-hanf-konnte-dem-bienensterben-entgegenwirken-DIEMEIGG3ZGHNJB-WDZ65RCQBQA.html (zugegriffen: 27.12.2021)

Pix, Reinhold (2020): Nutzhanf im Zeichen der Klimakrise, der nachhaltigen Landwirtschaft, der Rohstoffwende, in: Reinhold Pix Landtagsabgeordneter der Bündnis90/ Die Grünen, 29.01.2020, https://www.reinhold-pix.de/wp-content/uploads/2020/03/ Reader-Nutzhanf-02-2020.pdf (zugegriffen: 26.12.2021)

Plumb, Cedric (1941): Henry Ford's Hemp Car (1941): in: verymagazine, o. D., http:// www.verymagazine.org/magazine/216-overview-issue20/877-henry-fords-hemp-car-1941 (accessed: 26.12.2021)

Podbregar, Nadja (2020): Hanf, eine wassersparende Alternative zur Baumwolle?, in: Wissenschaft, 06.11.2020, https://www.wissenschaft.de/umwelt-natur/hanf-eine-wassersparende-alternative-zur-baumwolle/ (accessed: 26.12.2021)

Polizei (2020): Zahlen zu Drogendelikten, in: Polizei dein Partner Gewerkschaft der Polizei, 06.08.2020, https://www.polizei-dein-partner.de/themen/sucht/drogen/detailansicht-drogen/artikel/zahlen-zu-drogendelikten.html (accessed: 27.12.2021)

Porsche (2019): New Porsche 718 Cayman GT4 Clubsport featuring natural-fibre body parts, in: Porsche, 2019, https://www.porsche.com/international/aboutporsche/pressreleases/pag/?id=525217&pool=international-de&lang=none (accessed: 26.12.2021)

PottsAntiques (2010): The Hemp Car – Myth Busted, in: theangryhistorian, 24.10.2010, http://theangryhistorian.blogspot.com/2010/10/hemp-car-myth-busted.html (accessed: 26.12.2021)

Proplanta (2021): Agricultural subsidies 2020: Proplanta publishes list and top recipients. https://www.proplanta.de/agrar-nachrichten/unternehmen/agrarsubventionen-2020-proplanta-veroeffentlicht-liste-und-top-empfaenger_article1622025002.html (accessed: 26.12.2021)

Rätsch, Christian (2016): Hemp as a remedy, in: Google Books, 2016, https://books.google. de/books?hl=de&lr=&id=v494DwAAQBAJ&oi=fnd&pg=PT3&dq=hanf++Nahrung +rezepte&ots=F4Yi7vGWUA&sig=eyzMjQytDeBVFHHDBC8wu7mFNXw#v=onep age&q=hanf%20%20Nahrung%20rezepte&f=false (accessed: 25.12.2021)

Rehberg, Carina (2022): Hemp oil—one of the best cooking oils. https://www.zentrum-der-gesundheit.de/ernaehrung/lebensmittel/fette-oele-essig/hanfoel (accessed: 05.07.2022)

Richter, Susanne (2018): The cultivation of fiber hemp (Canabis sativa L.) as a winter cover crop, in: Bergische Universität Wuppertal, 2018, http://elpub.bib.uni-wuppertal.de/servlets/DerivateServlet/Derivate-8618/dd1807.pdf (accessed: 22.12.2021)

Riehl, Lisa (2019): How sustainable is the collection?, in: H&M Conscious Exclusive, 25.09.2019, https://www.harpersbazaar.de/nachhaltigkeit/hm-conscious-exclusive-kollektion-2019-nachhaltig (accessed: 26.12.2021)

Rinklebe, Jörg (2019): Final report on the project: Cultivation of hemp (Cannabis sativa L.) as a winter cover crop Reporting period from 20.07.2012 to 30.11.2016, in: Fachagentur Nachwachsende Rohstoffe, 09.04.2019, https://www.fnr-server.de/ftp/pdf/berichte/22015811.pdf (accessed: 26.12.2021)

Römer, Jörg (2019): Cement, the silent climate killer, in: DER SPIEGEL, Hamburg, Germany, 03.06.2019, https://www.spiegel.de/wissenschaft/zement-der-heimliche-klimakiller-a-0d863a07-d143-4335-a64e-b10de499af21?sara_ecid=soci_upd_wbMbjhOSvViISjc8RPU89NcCvtlFcJ (accessed: 26.12.2021)

Römer, Jörg (2021): 3D druck für Häuser, in: DER SPIEGEL, Hamburg, Germany, 02.10.2021, https://www.spiegel.de/wissenschaft/technik/3d-druck-fuer-haeuser-bautechnik-der-zukunft-a-22468972-4b73-46f9-9a34-1d5094ce5300?sara_ecid=soci_upd_wbMbjhOSvViISjc8RPU89NcCvtlFcJ (zugegriffen: 26.12.2021)

Rösemeier-Buhmann, Jürgen (2021): Monokultur: Wie der Anbau gleicher Sorten Landstriche zerstört, in: Nachhaltigleben.ch, 15.01.2021, https://www.nachhaltigleben.ch/food/monokultur-wie-eine-landwirtschaftsform-der-umwelt-schadet-2761 (zugegriffen: 26.12.2021)

Rolfsmeyer, Daniel (2017): Merkblatt Hanf Kulturanleitung Hanf (Cannabis sativa L.), in: Kompetenzzentrum Ökolandbau Niedersachen, 31.10.2017, https://www.oeko-komp.de/wp-content/uploads/2018/01/Merkblatt-Hanf.pdf (zugegriffen: 26.12.2021)

Rudorf, Julia (2021): Was können CBD-Produkte? In Apothekenrundschau (14.7.2021) https://www.apotheken-umschau.de/weitere-themen/was-bringen-cbd-produkte-777851.html (accessed: 05.07.2022)

Schildower Kreis (2015): Resolution of German criminal law professors to the members of the German Bundestag, in: Schildower Kreis, 25.10.2015, https://schildower-kreis.de/resolution-deutscher-strafrechtsprofessorinnen-und-professoren-an-die-abgeordneten-des-deutschen-bundestages/ (accessed: 27.12.2021)

Schöberl, Veronika; Fritz, Maendy; Grieb Michael (2019): Hemp for material use: Status and developments. Summary of the final report in: TFZ Straubing, 12.2019, https://www.tfz.bayern.de/mam/cms08/rohstoffpflanzen/dateien/191219_kurzfassung_hanfstoff_1107.pdf (accessed: 26.12.2021)

Schönberger, Hansgeorg and Pfeffer, Phillipp (2020): Agriculture: CO2-sinner or savior?, in: top agrar, 06.2020, https://www.topagrar.com/dl/3/7/4/3/1/6/2/CO2-Beitrag_final_Doppelseiten.pdf (accessed: 26.12.2021)

Schönthaler, Werner (o. J): Houses made of hemp. In: Build-Ing.: BIM-Fachmagazin und BIM-Plattform https://www.build-ing.de/fachartikel/detail/haeuser-aus-hanf/ (accessed: 16.11.2021)

Schönthaler, Werner (2022): Hanfstein I Hanfbeton: Eigenschaften. https://www.hanfstein.eu/home-deutsch/eigenschaften/ (accessed: 05.07.2022)

Schwager, Christian (2022): Das Geschäft mit Cannabis boomt – die Branche wartet auf die finale Freigabe (29.6.2022). https://www.berliner-zeitung.de/mensch-metropole/das-geschaeft-mit-cannabis-boomt-die-branche-wartet-auf-die-finale-freigabe-li.240881 (accessed: 05.07.2022)

Sensi Seeds (2020b): Alles über Hanffasern und die Hanf-Textilproduktion, in: Sensi Seeds, 30.04.2020, https://sensiseeds.com/de/blog/alles-uber-hanffasern-und-die-hanf-textilproduktion/ (accessed: 26.12.2021)

SGB 5 $ 31 – Einzelnorm, in: Gesetze im Internet, 2017, https://www.gesetze-im-internet.de/sgb_5/__31.html#:%7E:text=%C2%A7%2031%20Arznei%2D%20und%20Verbandmittel,Richtlinien%20nach%20%C2%A7%2092%20Abs.&text=6%20ausge-

schlossen%20sind%2C%20und%20auf,Verbandmitteln%2C%20Harn%2D%20und%20 Blutteststreifen (zugegriffen: 25.12.2021)

Statista (2020): Entwicklung des Cannabiskonsums unter Jugendlichen in Deutschland bis 2019, in: Statista, 27.07.2020, https://de.statista.com/statistik/daten/studie/219048/ umfrage/entwicklung-des-cannabiskonsums-unter-jugendlichen-in-deutschland/ (zugegriffen: 25.12.2021)

Statista (2020a): Statistisches Bundesamt: Rechtspflege Strafverfolgung Fachserie 10 Reihe 3, in: Destatis, 2020, https://www.destatis.de/DE/Themen/Staat/Justiz-Rechtspflege/Publikationen/Downloads-Strafverfolgung-Strafvollzug/strafverfolgung-2100300197004.pdf;jsessionid=E5249B881186FA3FCCEC174DC4D9624E. live721?__blob=publicationFile (zugegriffen: 27.12.2021)

Statista (2021): Bestand an zugelassenen Autos in Deutschland 2021, in: Statista, 08.09.2021, https://de.statista.com/statistik/daten/studie/12131/umfrage/pkw-bestand-in-deutschland/ (zugegriffen: 26.12.2021)

Steinort, Jennifer Ann: CBD Cosmetics Healthy Skin with Hemp, in: Krankenkassen-Zentrale, 01.08.2021, https://www.krankenkassenzentrale.de/produkt/cbd-kosmetik (accessed: 26.10.2021)

Stiftung Warentest (2021): Products with Hemp. Capsules and oils with CBD in the test: in: Stiftung Warentest, 25.02.2021, p. 86–91 (accessed: 05.07.2022)

Stiftung Warentest (2021a): Products with Hemp – What are capsules and oils with CBD worth, in: test.de, 26.01.2021, https://www.test.de/Produkte-mit-Hanf-Was-taugen-Kapseln-und-Oele-mit-CBD-5706119-0/ (accessed: 25.10.2021)

Stiftung Warentest (2021b): A coat for the house: thermal insulation: in:, Vol. 07, (2021, p. 64–67)

Stöver, Heino; Michels, Ingo; MüllerVahl, Kirsten; Grotenhermen, Franjo (2021): Cannabis as medicine: why further improvements are necessary and possible. In: akzept e. V., Deutsche Aidshilfe (ed.) 8. Alternative Drugs and Addiction Report 2021 (Akzept/Aidshilfe 2021, p. 142–148)

Suliak, Hasso (2022): Planned cannabis legalization What comes into the law of the traffic light? LTO Legal Tribune Online 1.7.2022. https://www.lto.de/recht/hintergruende/h/cannabis-legalisierung-ampel-konsultation-gesetzentwurf-jugendschutz-richtervorlage-blienert/ (accessed: 05.07.2022)

Suman, Chandra; H. Lata; ElSohly (eds) (2017): Cannabis sativa L. – Botany and Biotechnology Springer

Telgheder, Maike (2021a): Cannabis „Made in Germany": Domestic hemp conquers pharmacies, in: Handelsblatt, 07.07.2021, https://www.handelsblatt.com/unternehmen/industrie/marihuana-als-medizin-made-in-germany-jetzt-erobert-heimisches-hanf-die-apotheken/27393830.html?ticket=ST-5574046-ta6oELCRpXG115FIcytJ-cas01.example.org (accessed: 22.12.2021)

Telgheder, Maike (2021a): Cannabis as medicine: weaker growth than expected, in: Handelsblatt, 19.01.2021b, https://www.handelsblatt.com/unternehmen/industrie/marihuana-als-medizin-zahl-der-cannabis-patienten-steigt-aber-nicht-so-schnell-wie-erwartet/26793480.html?ticket=ST-6880985-woRB5YAZLy5CWVVGzgwO-cas01. example.org (accessed: 25.12.2021)

Tietjen, Daniel; Behm, Christoph (2021): Unity of the coalition of traffic lights with regard to the legalization of cannabis products. First details in the new coalition agreement. In:

https://www.taylorwessing.com/de/insights-and-events/insights/2021/12/einigkeit-der-ampel-koalition-hinsichtlich-der-legalisierung-von-cannabis-produkten   (zugegriffen: 05.07.2022)

Tilray (o. J.): Deutschland GmbH: Die Zukunft des medizinischen Cannabis. Gemeinsam gestalten, in: Tilray, o. D., https://tilray.de/ (zugegriffen: 26.12.2021)

Traufetter, Gerald (2021): Greenpeace und Deutsche Umwelthilfe leiten Klage gegen deutsche Großkonzerne ein. In: Der Spiegel Nr. 36, 2021. https://www.spiegel.de/wirtschaft/unternehmen/klimakrise-greenpeace-und-deutsche-umwelthilfe-leiten-klage-gegen-deutsche-grosskonzerne-ein-a-a83a69fe-3035-46d5-9f08-dd157192b5dc (zugegriffen: 05.07.2022)

Umweltbundesamt (2014): Der Weg zum klimaneutralen Gebäudebestand https://www.umweltbundesamt.de/sites/default/files/medien/378/publikationen/hgp_gebaeudesanierung_final_04.11.2014.pdf (zugegriffen: 26.12.2021)

Federal Environment Agency (2015): Microplastics in the sea—how much? Where from?, in: Federal Environment Agency, 29.09.2015, https://www.umweltbundesamt.de/presse/pressemitteilungen/mikroplastik-im-meer-wie-viel-woher (accessed: 26.12.2021)

Federal Environment Agency (2018): Evaluation of the judgment of the European Court of Justice (ECJ) of 21 June 2018 in Case C-543/16 (Commission against the Federal Republic of Germany) for breach of contract (Nitrates Directive 91/676/EEC), in: Federal Environment Agency, 27.06.2018, https://www.umweltbundesamt.de/sites/default/files/medien/2875/dokumente/uba-auswertung_eugh_urteil_2018-07-26.pdf (accessed: 26.12.2021)

Federal Environment Agency (2019): Living and renovating. Empirical residential building data since 2002 https://www.umweltbundesamt.de/sites/default/files/medien/1410/publikationen/2019-05-23_cc_22-2019_wohnenundsanieren_hintergrundbericht.pdf (accessed: 05.07.2022)

Federal Environment Agency (2020): Global car stock, in: Federal Environment Agency, 2020,   https://www.umweltbundesamt.de/bild/weltweiter-autobestand   (accessed: 26.12.2021)

Federal Environment Agency (2022): IPCC report: Immediate global trend reversal necessary. ()13.5.2022) https://www.umweltbundesamt.de/themen/ipcc-bericht-sofortige-globale-trendwende-noetig (accessed: 05.07.2022)

University of Vienna (2011): Calculation of carbon fixation in reforestation in the tropics, in: Regenwald, 01.12.2011, https://www.regenwald.at/fileadmin/content/filebrowser/PDF_Dokumente/CO2_Berechnung_Uni.pdf (accessed: 26.12.2021)

Unkart, Enya (2020): Hemp seeds: ingredients, effects and applications, in: Utopia.de, 16.09.2020, https://utopia.de/ratgeber/hanfsamen-inhaltsstoffe-wirkung-und-anwendung (accessed: 22.12.2021)

UNODC (o. J.): United Nations Office on Drugs and Crime: International Drug Control Conventions. https://www.unodc.org/unodc/en/commissions/CND/Mandate_Functions/Mandate-and-Functions_Scheduling.html (accessed: 27.12.2021)

USDA (2019): National Nutrient Database for Standard Reference: Hemp seed, in: USDA, 04.09.2019, https://www.uwyo.edu/ipm/_files/docs/ag-ipm-docs/hemp-ipm-docs/usda-full-nutrient-report.pdf (accessed: 25.12.2021)

United Nations (2016): SDG Report 2016, in: United Nations Sustainable Development Goals, 01.01.2016, https://www.un.org/depts/german/millennium/SDG%20Bericht%20 2016.pdf (accessed: 14.10.2021)

United Nations (2020): SDG Report 2020, in: United Nations Sustainable Development Goals, 01.01.2020, https://www.un.org/Depts/german/millennium/SDG%20Bericht%20 2020.pdf (accessed: 15.10.2021)

Berlin Administrative Court (2021): No distribution of CBD products without testing (No. 13/2021): in: Berlin Press Release, 15.03.2021, https://www.berlin.de/gerichte/verwaltungsgericht/presse/pressemitteilungen/2021/pres-semitteilung.1064355.php (accessed: 25.10.2021)

Vosper, James (2020): The Role of Industrial Hemp in Carbon Farming, in: Parliament of Australia, 05.06.2020, https://www.aph.gov.au/Help/Federated_Search_Results?q= The+Role+of+Industrial+Hemp+in+Carbon+Farming&ps=10&pg=1 (zugegriffen: 26.12.2021)

Walch-Nasseri, Friederike (2022): IPBES World Biodiversity Council: Without wilderness, man dies. An analysis, 12.7.2022, https://www.zeit.de/wissen/umwelt/2022-07/weltbio-diversitaetsrat-ipbes-artenschutz-bericht-umweltschutz (zugegriffen; 12.7.2022)

Welt der Wunder (2021): Plastics Alternatives in Test—More Show than Being?, in: Welt der Wunder TV, 24.05.2021, https://www.weltderwunder.de/artikel/plastik-alternativen-im-test-mehr-schein-als-sein (zugegriffen: 26.12.2021)

Wieland, Hansjörg; Bockisch, Franz-Josef (2003): German Sheep's Wool—Insu-lation with a Future?, in: Landtechnik-Online, 2003, https://web.archive.org/ web/20180521053718id_/https://www.landtechnik-online.eu/ojs-2.4.5/index.php/land-technik/article/viewFile/2003-4-260-261/2642 (zugegriffen: 26.12.2021)

Winkler, Gillian (2019): 5 Millionen Euro: Der größte Cannabis-Transport I Galileo I ProS-ieben, in: YouTube, 19.11.2019, https://www.youtube.com/watch?v=zSjYM3p9lKU (zugegriffen: 26.12.2021)

Wir leben nachhaltig (o. J.): Hanf ist eine vielseitige Pflanzehttps://www.wir-leben-nachhaltig.at/aktuell/detailansicht/lebensmittel-und-superfood-hanf (zugegriffen: 25.12.2021)

Wohlers, Katja (2019): Verordnung: Was ist zu beachten?, in: Die Techniker, 11.12.2019, https://www.tk.de/techniker/gesundheit-und-medizin/behandlungen-und-medizin/can-nabis-verordnung-was-beachten-2032620?tkcm=ab (zugegriffen: 25.12.2021)

Wurth, Georg (2020): CBD: Stellvertreterkrieg um Cannabidiol In: akzept e. V., Deutsche Aidshilfe (Hrsg.) 7. Alternativer Drogen- und Suchtbericht. Pabst Science Publishers (Lengerich) (2020 S. 149–156)

WVCA (o. J.): Wirtschaftsverband Cannabis Austria: Über uns – WVCA, https://www. wvca.at/ueberwvca (zugegriffen: 26.12.2021)

WWF (2019): Climate protection in the concrete and cement industry. Background and options for action https://www.wwf.de/fileadmin/fm-wwf/Publikationen-PDF/WWF_ Klimaschutz_in_der_Beton-_und_Zementindustrie_WEB.pdf (accessed: 05.07.2022)

Center for Health (https://www.zentrum-der-gesundheit.de/ernaehrung/lebensmittel/fette-oele-essig/hanfoel (accessed: 05.07.2022)

Printed in the United States
by Baker & Taylor Publisher Services